Also by Mostafa Dolatyar

THE LEGAL REGIME OF INTERNATIONAL STRAITS (*in Persian*)

Also by Tim S. Gray

BURKE'S DRAMATIC THEORY OF POLITICS (*co-author*)
THE FEMINISM OF FLORA TRISTAN (*co-author*)
FREEDOM
INDIVIDUALISM AND ORGANICISM
THE POLITICAL PHILOSOPHY OF HERBERT SPENCER
THE POLITICS OF GENETIC RESOURCE CONTROL (*co-author*)
THE POLITICS OF FISHING (*editor*)
UK ENVIRONMENTAL POLICY IN THE 1990s (*editor*)

Water Politics in the Middle East

Water Politics in the Middle East

A Context for Conflict or Co-operation?

Mostafa Dolatyar
Ministry of Foreign Affairs
Iran

and

Tim S. Gray
Professor of Political Thought
Department of Politics
University of Newcastle
Newcastle upon Tyne

First published in Great Britain 2000 by
MACMILLAN PRESS LTD
Houndmills, Basingstoke, Hampshire RG21 6XS and London
Companies and representatives throughout the world

A catalogue record for this book is available from the British Library.

ISBN 0-333-74503-5

First published in the United States of America 2000 by
ST. MARTIN'S PRESS, INC.,
Scholarly and Reference Division,
175 Fifth Avenue, New York, N.Y. 10010

ISBN 0-312-22382-X

Library of Congress Cataloging-in-Publication Data
Dolatyar, Mostafa.
Water politics in the Middle East : a context for conflict or
co-operation? / Mostafa Dolatyar and Tim S. Gray
p. cm.
Includes bibliographical references and index.
ISBN 0-312-22382-X (cloth)
1. Water-supply—Middle East. 2. Water-supply—Political aspects-
—Middle East. 3. Water resources development—Middle East.
4. Middle East—Strategic aspects. I. Gray, Tim, 1942– .
II. Title.
HD1698.M53D65   1999
333.91'00956—dc21   99–22104
                                                                 CIP

© Mostafa Dolatyar and Tim S. Gray 2000

All rights reserved. No reproduction, copy or transmission of this publication may be made without written permission.

No paragraph of this publication may be reproduced, copied or transmitted save with written permission or in accordance with the provisions of the Copyright, Designs and Patents Act 1988, or under the terms of any licence permitting limited copying issued by the Copyright Licensing Agency, 90 Tottenham Court Road, London W1P 0LP.

Any person who does any unauthorised act in relation to this publication may be liable to criminal prosecution and civil claims for damages.

The authors have asserted their rights to be identified as the authors of this work in accordance with the Copyright, Designs and Patents Act 1988.

This book is printed on paper suitable for recycling and made from fully managed and sustained forest sources.

10 9 8 7 6 5 4 3 2
09 08 07 06 05 04 03 02 01 00

Printed and bound in Great Britain by
Antony Rowe Ltd, Chippenham, Wiltshire

Contents

List of Maps	viii
List of Tables and Figures	ix
Preface	x
Acknowledgements and Dedication	xii
List of Abbreviations	xiii

1 Introduction   1

The importance of the topic	1
Water, the core of civilisation	1
Water, the principal resource problem of the world	5
The scope of the study	8
Water in the Middle East	8
The framework of analysis	10
Organisation of the chapters	12
Notes	13

2 Five Approaches to Water Scarcity   15

Introduction	15
Five hypotheses about the water issue	18
Water as a security issue	18
Water as an economic issue	23
Water as a legal issue	32
Water as a technical issue	40
Water as an environmental issue	47
Conclusion	56
Notes	57

3 Water Scarcity: A Global Problem?   60

Water management: a global challenge	60
Conceptualisation	61
What is water scarcity?	61
Is it a global problem?	62

vi *Contents*

Global water demand	65
Regional water problems	67
Europe	67
America	69
Asia	72
Middle East	79
Conclusion	82
Notes	83

4 Water Politics in the Jordan River Basin   **85**

Introduction	85
Climatic and hydropolitical features	88
Evolution of the water crisis in the Jordan Basin – the role of Zionism	94
Historical background	94
Early Zionism	96
After independence	98
Israel's water policy and Arab reaction	103
Conception of water as a strategic issue	103
Institutionalisation of water policies	104
The new paradigm	108
Conclusion	112
Notes	114

5 Water Politics in the Euphrates–Tigris Basin   **116**

Introduction	116
Climatic and hydropolitical features	119
Evolution of water management in Mesopotamia	124
Ancient Mesopotamia	124
Modern Mesopotamia	130
Turkey's water policy and Arab reaction	143
Conception of water as a strategic issue	143
Water diplomacy	146
Conclusion	158
Notes	161

6 Water Politics in the Arabian Peninsula   **164**

Introduction	164
Climatic and hydropolitical features	167

Contents vii

Evolution of water management in the Arabian Peninsula 171
  Ancient Arabia 171
  Modern Arabia 178
  Water management policies of the regional states 185
  Groundwater management 186
  Non-conventional water management 188
  The dawn of a new paradigm 198
  Conclusion 200
  Notes 203

7 **Conclusion** **206**

  Summary 206
  Two Hypotheses 208
  Hypothesis 1 – not conflict, but co-operation, in the past 209
  Hypothesis 2: the new paradigm – environmentalism –
  for the future 217
  Notes 219

Bibliography 220

Index 244

List of Maps

1.1 The ancient hydraulic civilisations of the Middle East 4
4.1 The Jordan River basin 88
4.2 The Hullah wetlands 101
5.1 The Euphrates–Tigris basin 120
5.2 The 'Third River' in Iraq 159
6.1 Countries of the Arabian Peninsula 166
6.2 Shared international aquifers of the Arabian Peninsula 169
6.3 The 'Peace Pipeline' and the Iran–Qatar water project 196

List of Tables and Figures

Tables

2.1	Global water volumes	24
3.1	International river basins with more than four riparian countries	64
3.2	Water use by continent (bcm/year)	65
3.3	The share of agriculture in national economy and employment in Central Asian republics	79
3.4	Per capita water availability and the ratio of its demand to supply in the Middle East	81
4.1	Comparative hydrology of the major rivers in the Middle East	87
4.2	Major tributaries of the Jordan River	90
4.3	Water supply and demand in the Jordan River basin	91
4.4	Jewish population growth in Palestine/Israel	99
4.5	Agricultural development in Israel	106
4.6	The position of agriculture in the economic structure of Israel	109
5.1	A partial chronology of conflict over water in the ancient Middle East	128
5.2	Per capita surface water availability in the Euphrates–Tigris basin	140
5.3	Annual surface water and withdrawal in the Euphrates–Tigris basin	141
6.1	Water balance in the Arabian Peninsula	170
6.2	Population growth in the Arabian Peninsula (millions)	180
6.3	Water deficit in the Arabian Peninsula (millions m^3/yr)	183
6.4	Per capita water availability from natural sources (1990)	184
6.5	Wastewater treatment and reuse in the oil-rich countries of the Arabian Peninsula	193
6.6	The 'Peace Pipeline' capacity and delivery points	197

Figures

2.1	Technological options for managing water scarcity	42
2.2	Managerial options for managing water scarcity	44
6.1	The structure of a *qanat* system	177

Preface

Most recent studies and reports describe a grim picture of fresh water availability in the Middle East, indicating that there is a significant risk of imminent *conflicts* and *wars* over water in this region. Apart from exaggerating the scarcity of water resources and the likelihood of war, the major weakness in this literature is that it is more problem-orientated than solution-orientated.

This study is an inquiry into the conduct of riparian states in transnational river basins of the Middle East, based on an analysis of the actual needs of the countries bordering these basins and the political implications of the unevenly distributed region's water resources. It seeks to find out the obstacles which have prevented the countries of the region reaching a cooperative basin-wide arrangement, which is the optimal method for development and exploitation of their common water resources. The scope of concern includes both the transnational rivers and the cross-border aquifers in the Middle East where, because of the aridity of the climate and the high rate of population growth, unimpeded access to freshwater resources is linked to national survival.

The significance of this study lies in its endeavour to define the limitations and opportunities for the achievement of cooperative solutions to the problem of managing a common property resource and to avoid both the 'tragedy of the commons' and regional violence. The main objective is to put forward an interpretation of water politics in which water is seen as a critical factor in moving countries toward cooperation rather than military conflict with their co-riparian neighbours. It will show that, although water has occasionally provoked dispute in the Middle East, it has much more often promoted coexistence between adversaries. The main hypothesis is that, contrary to the most frequently mentioned scenario in the literature that suggests that dispute over water supplies will lead to interstate war, it is unlikely that the quest for more water will cause a new war in the Middle East. Rather, water shortage should be seen as a platform for regional cooperation that promises development and exploitation of the region's water supplies in ways that all riparian nations can achieve optimal solutions. Moreover, joint cooperative development of common water resources will actually reinforce peace. The reasoning behind this hypothesis is that:

> water is too vital to be deliberately put at risk by war because due to the aridity of the region, the acute needs of the region's nations, and the vulnerability of the expensive water engineering systems such as dams, hydropower installations, and water distribution networks, it is a mutual hostage situation and all parties are at risk, should there be any war. (Deudney 1991: 26)

The fact is, as Peres with Naor (1993: 128) have rightly acknowledged, 'wars fought over water do not solve anything. Gunfire will not drill wells to irrigate the thirsty land, and after the dust of war has settled, the original problems remain.' Viewed in this light, the water shortage reduces rather than increases the likelihood of states in the region employing violence to resolve the problem. Indeed, there is evidence that it encourages the joint development of common water resources based on a network of common interests. Alam calls this 'water rationality' – the belief that 'co-operation over international water resources is more probable than conflict' (Alam 1998: 251).

MOSTAFA DOLATYAR
TIM S. GRAY

Acknowledgements and Dedication

This book is based on a doctoral thesis submitted by Mostafa Dolatyar in 1998 under the supervision of Tim Gray at the Politics Department of the University of Newcastle upon Tyne. We would like to thank both the Foreign Ministry of the Islamic Republic of Iran for the award of a scholarship that enabled Mostafa to study in Newcastle, and Newcastle University for the resources that facilitated the completion of the project.

The book is dedicated to Mostafa's wife, Sharareh SheikBahaie, without whose selfless support and encouragement it would never have appeared.

M. D.
T. S. G.

List of Abbreviations

ADB	African Development Bank
AOAD	Arab Organization for Agricultural Development
APEC	Asia Pacific Economic Cooperation
CIA	Central Intelligence Agency
EIB	European Investment Bank
EOSAT	Earth Observation Satellite Company
EPA	Environmental Protection Agency (USA)
EU	European Union
FAO	Food and Agriculture Organization
GAP	Güneydogu Anadolu Projesi (Turkish Water Management Scheme on the Euphrates River)
GIS	Geographical Information System
GRAND	Great Recycling and Northern Development (Canada)
ICWE	International Conference on Water and Environment (Dublin 1992)
ILA	International Law Association
ILC	International Law Commission
ILO	International Labour Organization
IMO	International Maritime Organization
INWC	Israeli National Water Carrier
JNF	Jewish National Fund
NATO	North Atlantic Treaty Organization
NAWAPA	North-American Water and Power Alliance
OECD	Organization for Economic Cooperation and Development
OAU	Organization of African Unity
[P]GCC	[Persian] Gulf Co-operation Council
PKK	Kurdistan Workers' Party
PNLM	Palestinian National Liberation Movement (Al-Fatah)
SIPRI	Stockholm International Peace Research Institute
UNCED	United Nations Conference on Environment and Development (Rio de Janeiro 1992)
UNCLNUIW	United Nations Convention on the Law of the Non-navigational Uses of International Watercourses (1997)
UNCLOS	United Nations Convention on the Law of the Sea (1982)

UNDP	United Nations Development Programme
UNECE	United Nations Economic Commission for Europe
UNEP	United Nations Environment Programme
UNESCO	United Nations Education, Scientific and Cultural Organization
UNGA	United Nations General Assembly
UNWC	United Nations Water Conference (Mar del Plata 1977)
WCED	World Commission on Environment and Development (Brundtland Commission)
WHO	World Health Organization

1
Introduction

The importance of the topic

Water, the core of civilisation

Water as the sum and substance of life has decisively affected the course of human history. Its abundance has enabled societies to flourish, its scarcity has caused them to wither. No consideration of the fate of societies can ignore water's determining role. Great civilisations have grown up along great rivers. Allured by the life-sustaining waters, different races of humankind have migrated to the banks of the rivers and oases; and as settlements were developed, water abundance or scarcity has deeply influenced their culture, religion, politics, economy, and their lifestyle.

In the water-rich temperate regions, people took water for granted. They perceived it as a free gift of nature whose economic and efficient use was not a major concern. Indeed, the abundance of water relative to demand engendered a state of complacency and the failure to appreciate the fact that water is a finite resource. In the water-poor arid regions, by contrast, people have always revered water as a finite resource which is to be conserved, protected, and respected as a giver of life. In Arabic culture, for instance, which sprang out of life in the desert, water is deemed sacred and its waste or pollution considered a sin. In cultures developed in arid and semi-arid regions water is regarded not only as a physical agent but also as a source of spiritual purification and renewal. Persian literature, for example, is filled with delighted, sensitive accounts of the coming of rain, a momentous event. The first word in the Persian dictionary is *äb* (water) from which the words *äbad* (habitable) and *äbadan* (civilised) are derived.

2 *Water Politics in the Middle East*

The core of Mecca, the holiest city of Islam, which non-Muslims are not permitted to enter, surrounds a great mosque. Inside the mosque, the *Ka'bah* (a shrine) and the *Zamzam* (a well of fresh water) are located. They are the focal point of the pilgrimage. Similarly, the Ganges River, the chief river in India which flows through the heart of the nation, has a religious meaning to hundreds of millions of people. There are several holy sites along its banks, including Varanasi (Benares), the holiest Hindu city with some 1500 temples, palaces, and shrines. About 1 million pilgrims annually visit the city to bathe in the sacred waters of the Ganges. Buddhism was also born in the Ganges River basin, where Siddhartha Gautama (Buddha) travelled up and down the river and founded monastic orders of a religion that became dominant in China, Japan, and other Asian countries.

The Middle East is the cradle of the earliest known civilisations of the ancient world and the birthplace of Judaism, Christianity, and Islam. In this region, wide seasonal temperature variations and highly irregular rainfall affected how the ancient Israelites and Arabs viewed themselves and their God. They conceived themselves entirely dependent on God to send rain in its seasons. The blessing of rain was presumed to be contingent on their faithfulness in their relationship with God. People, however, managed to transform the 'hydrological chaos' of the region into the well-organised and regularly watered fields and gardens such as the Hanging Gardens of Babylon, one of the Seven Wonders of the World. Their continuing endeavour to 'make the desert bloom' represents people's desire and enthusiasm for realising the Garden of Eden on earth.

As described in the Koran (2: 35; 7: 19), as well as in the Bible (Genesis ch. 2), humanity's first habitat was the Garden of Eden. In it was grown 'every tree that is pleasant to the sight and good for food', all watered by a perpetual river and its distributaries. Adam and Eve were put in the garden 'to serve and keep it' (Genesis 2: 15). Unfortunately, however, according to both the Koranic and the biblical stories, the first humans soon abused God's trust. Yielding to enticement and consuming beyond their needs, they cast themselves out from the gardens of bliss and eternity. The followers of these three great monotheistic religions have been living with the desire of a return to the Paradise of Eden. But only the faithful are promised eternal life in the Paradise which is described in the Koran as 'gardens of everlasting bliss beneath whose trees perpetual incorruptible rivers flow' (5: 119), in which 'every so often they are fed with fruits therefrom' (2: 25), and admitted to 'shades, cool and ever deepening' (4: 57). Undoubtedly, to inhabitants of a hot and dry climate, such statements portray the ideal of serenity, joy, and well-being.

In the Judeo-Christian and Muslim traditions, there are numerous direct references to rivers and springs in Paradise. Indeed, 'running waters' are an essential component of the concept of Paradise. The Hebrew word for heaven, *shamayim*, can be separated into *shammayim*, which suggests 'source of water' (Hillel 1994: 24; Starr 1995: 15). Moreover, water is a major theme in the Koran, with more than 150 references to water, rivers, and springs. The Koran repeatedly reminds us that water is the essence of creation, life, and joy. Allah's own throne is on the water (11: 9), and near a river or a pool in the Paradise called *Kauthar*, all Muslims believe that the Prophet will stand to greet the faithful.

Springs and wells have also been the focal points of social life, including disputes, in both nomadic and settled societies of the Middle East. One of the earliest references to a dispute over water is given in Genesis (21: 25–32), wherein 'Abraham reproved Abimelech (king of the Philistines) because of the well of water which Abimelech's servants had violently taken away' from Abraham's servants. To settle the dispute, 'Abraham set seven ewe lambs ... and he said, for these seven ewe lambs shalt thou take of my hand, that they may be a witness unto me, that I have digged this well. ... Thus they made a covenant'. This episode illustrates that apportionment of the scarce water resources has sometimes been a matter of dispute. Significantly, however, it shows also that the response to disputes has not been inflammatory but thoughtful and sensible. While developing their own water resources, the parties engaged in meaningful negotiations to resolve potential conflicts in an amicable fashion.

Hydraulic civilisations

Although the theories concerning the determination of societies and their cultures by infrastructural factors are highly contested, the far-reaching influence of climatic and hydrological conditions on early agrarian civilisations so convinced the German historian and anthropologist Karl Wittfogel (1956) that he introduced the notion of 'hydraulic civilisations'. In his classic work *Oriental Despotism: a Comparative Study of Total Power*, Wittfogel formulated the thesis that the need to develop great infrastructures for irrigation

Map 1.1 The ancient hydraulic civilisations of the Middle East

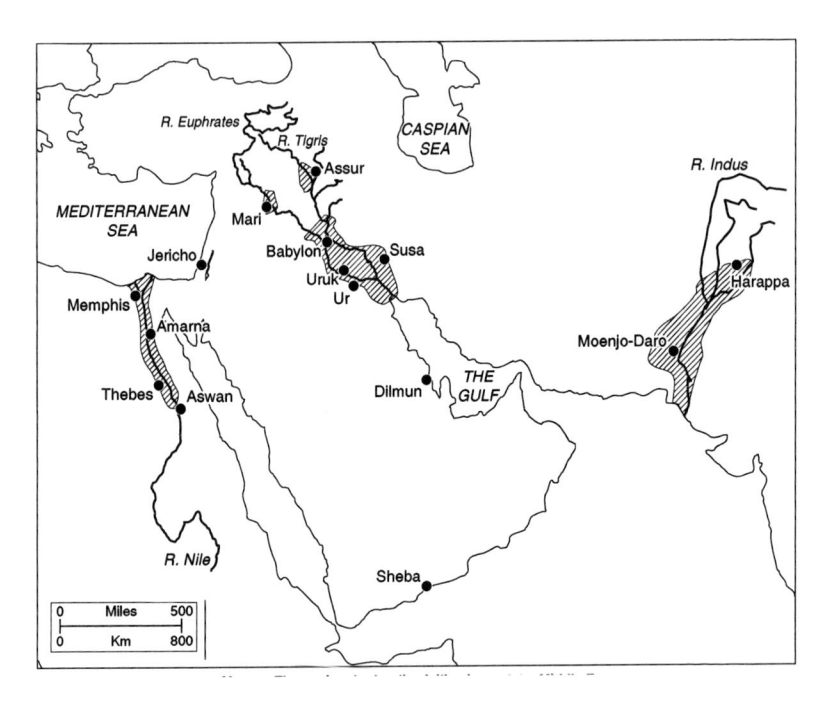

Source: Hillel (1994: 52) with minor modifications

has been the incentive for building bureaucratic organisations, from which emerged the first forms of centralist states. Prime examples are the civilisations of the Euphrates–Tigris, the Nile, and the Indus rivers (see Map 1.1).

Historically, in most parts of these regions, permanent settlements were only possible along rivers which provided the water needed for irrigation. The challenge to control the extreme fluctuation of the rivers and make efficient use of water resources played an important role in the emergence of large-scale infrastructures and organised institutions. Having observed this historical pattern, Wittfogel wrote that in arid or semi-arid environments 'agrarian civilisations can persist permanently and prosperously only on the basis of hydraulic economy'. Highlighting the basic characteristics of hydraulic economy, he argued that hydraulic agriculture involves a specific type of division of labour, intensive cultivation, and large-scale cooperation. Therefore, a state is needed which is stronger than society (Wittfogel 1956: ch. 3). 'This is despotic power, total and non-benevolent.... The despotic character of hydraulic government is not seriously contested' (Ibid.: 101). Wittfogel was right in his conception of water sharing as a central principle to Oriental societies. He might also be right to claim that in past civilisations rulers were able to use their control of irrigation networks as a means to political power. However, establishing a general theory on the ground that societies which rely critically on irrigation systems are characterised by rigidly centralised control is another matter. Wittfogel did not appear to recognise that there has been a rich legal tradition and a legacy of civil institutions in some of these societies, which were not despotic as he presumed. For instance, the institutionalisation of an exemplary code of laws such as the Code of Hammurabi, which in the eighteenth century BCE established a systematic set of laws and rules pertaining to issues such as family, labour, personal property, real estate, trade, and business transactions, including rules for maintaining an efficient water management system, indicates that the Mesopotamian civilisations had well-established civil institutions to manage their society.[1] These traditions were considerably enriched and developed in the Islamic era.[2]

Water has always been central to the Islamic world. It has been central not only in obvious terms of scarcity in arid or semi-arid land, but also in the sense that water lay at the heart of the legal system. Indeed, before it simply meant *law*, the Arabic word for Islamic law, *shari'a*, denoted the law of water. Ibn Manzur (d. 1311 AC), the famous Arab lexicographer, mentions in his dictionary *Lisan al-'Arab* (vol. 3: 175) that '*shari'a* is the place from which one descends to water ... and *shari'a* in the acceptation of Arabs is the law of water (*shur'at al-ma'*) concerning the source which is regulated by people who drink, and allow others to drink, from.' The same word is also used for what God has decreed through revelation. The connection between *shari'a* as the law of water and as a generic term for Islamic law is not a matter of coincidence. As mentioned above, the centrality of water in Islam is evident in both an economic and a ritualistic sense. What is more important is that the jurists – the exponents and expounders of the *shari'a* – developed and codified, in answer to this centrality, an extensive and highly sophisticated system of rules which is far from being despotic (see Chibli Mallat 1995: 127–37). Unfortunately, however, this rich and refined literature is almost always neglected by Western observers who comment on water resources management in the Middle East.

Water, the principal resource problem of the world

As the essence of life and a prime resource on which our global society is based, water has always been the principal challenge for humanity from the early days of civilisation. This challenge is shaped by the fact that, unlike most resources, there is no alternative to water. Along with population growth and socio-economic development, the world's total water consumption has surged 35-fold during the past 300 years, and more than half of that increase has occurred since 1950 (Abramovitz 1996: 31). This rate of increased water consumption is not expected to decline, yet today's global per capita water availability is predicted to fall by a third over the next generation (World Bank 1995: 5). Furthermore, it is not just the quantity of available water which is a matter of concern. Many of the world's rivers, lakes, and aquifers, serving as sources of fresh water, have become so polluted with toxic materials or contaminated by pathogens that they are no longer safe for human use (see Chapter 3).

It is important to note, however, that water resource development is not just the domain of engineers. Rather, as Falkenmark (1990: 189) points out, it presents a multitude of complex management challenges derived from the interaction between human behaviour and the water cycle. It is human behaviour in the living environment, in particular humankind's way of living with available water, and of accepting the integrity of the water cycle and the laws which control the water cycle, that determines the balance between supply and demand. When water resource management is handled properly, it can provide the basis for economic growth, improvements in living standards and sociopolitical stability. The experience of the past decades, however, shows that these benefits are undermined by poor management of water resources.

Uncertainties regarding population growth, agricultural needs, industrial requirements, and possible global warming have made long-term assessments of water resources difficult. However, we can be sure that water use will continue to increase; water pollution will continue to put pressure on usable water supplies; and water will continue to be the principal resource problem of the world. Indeed, many studies, reports, and assessments suggest that a looming 'world-wide water crisis' is on the horizon.[3] These studies indicate that present and expanding water uses will increase transnational dependencies on shared water resources in different parts of the world. In fact, the extent of water interdependency is already very extensive. Over 240 river basins are shared between two or more states (Blake *et al.* 1995: xiv), comprising about 50 per cent of the land and more than 40 per cent of the world's population (Vlachos *et al.* 1986: 1–2). The management and development of these shared water resources pose special challenges, which sometimes turn into explosive political issues. As the demand for water increases, and as exclusively indigenous sources of water are fully developed, the only other sources are likely to be international. But, in the absence of explicit regulations, international sources often breed tension. For example, in a river flow shared by several countries, unchecked upstream developments could prove a matter of serious concern to downstream users. This is why water security is already one of the most crucial elements in the foreign policy considerations of many countries.

Apprehension of this global trend towards increased pressure on water resources has elevated the issue of water resources management to a new diplomatic level. It was the increasing realisation of the importance of water in the continuing well-being and development of humankind and its effects on international relations that prepared the ground for the United Nations Water Conference in Mar del Plata, Argentina, in March 1977. The conference's primary purpose was to promote a level of preparedness that would enable the world to avoid a water crisis of global dimensions by the end of the present century (UNWC 1977). Since that conference, in almost all countries of the world, including advanced industrialised and water-rich countries, water availability and its rational management has become a major sociopolitical issue. The International Conference on Water and Environment (ICWE) in Dublin (January 1992), drafted an action plan for the water sector which was submitted to the UN Conference on Environment and Development (UNCED) in Rio de Janeiro (June 1992). Chapter 18 of the UNCED's Agenda 21 formulated a comprehensive action plan devoted to water issues, whose 'general objective is to make certain that adequate supplies of water of good quality are maintained for the entire population of this planet, while preserving the hydrological, biological and chemical functions of ecosystems' (UNCED 1992, ch. 18, para. 2).

Observers outside the diplomatic circle have also drawn attention to the seriousness of the issue at stake. Many of them paint a bleak picture of a world without enough water (see Chapter 3). Falkenmark (1990: 183), for example, believes that 'the serious water problems now confronting humanity are all inherently international and many of them even global in scope'. Clarke (1991: 169, 178) argues that 'water ranks alongside food and energy as a key global resource – if not above them, ... and managing water better – locally, nationally and internationally – is now a matter of life and death'. Winpenny (1994: 2) claims that 'water is becoming one of the largest, and certainly the most universal, of problems facing mankind as the earth moves into the twenty-first century'. Ismail Serageldin, the World Bank vice-president for environmentally sustainable development, believes that 'many of the wars this century were about oil, but wars of the next century will be over water' (*Financial Times*, 7 August 1995: 2).

The scope of the study

Water in the Middle East

While it is clear that, both in terms of quality and quantity, water scarcity is and will be a major issue in several regions of the world, agencies such as the Intelligence Agency of the US Department of Defence (Naff and Matson 1984), the Center for Strategic and International Studies (Starr and Stoll 1987), the Gulf Centre for Strategic Studies (Musallam 1990), and the International Institute for Strategic Studies (Beschorner 1992) have especially highlighted the issue in the Middle East. Many writers and observers from varied backgrounds have also focused on the magnitude of the water problem in the Middle East,[4] and some leading journals have devoted entire issues to the topic of Middle Eastern hydropolitics.[5] Naff and Matson (1984: 3), for instance, have maintained that it is in the Middle East that 'water-generated' conflicts could easily engulf the entire region. Likewise, Starr and Stoll (1987) in a report entitled *U.S. Foreign Policy on Water Resources in the Middle East*, have predicted that water, not oil, will become the dominant subject of conflict for the region by the year 2000. During the mid-1980s, Starr (1991: 17) has revealed, US intelligence services estimated that 'there were at least 10 places in the world where war could break out over dwindling shared water – the majority in the Middle East'.

For many analysts, the water issue is the epitome of the Middle East conflict. Defining the Middle East as 'a region of international concern and political unrest with severe water shortages', Kliot (1994:

Introduction 9

V) reiterates the claim that, 'water, not oil, threatens the renewal of military conflicts and social and economic disruption in the region'. Such assumptions led Starr (1991: 19) to pronounce that in the Middle East 'water security soon will rank with military security in the war rooms of defence ministries'. Indeed, the hydropolitics literature is replete with assertions such as Butler's (1995: 34) that 'nowhere is the potential for a water war greater than in the Middle East'. (See for example: Musallam 1990; Postel 1992: 29, 74; Bulloch 1993; Bulloch and Darwish 1993; Starr 1993, 1995; Hillel 1994; Lonergan 1997: 418.) Almost all these studies and reports describe a grim picture of fresh water availability in the Middle East and indicate that there is a big risk of coming conflicts[6] and wars[7] over water in this region.

These assertions and hypotheses will be discussed fully in subsequent chapters. However, it is important to state here that the main concern of this study is to establish that the conventional idea of 'water war in the Middle East' has become something of a cliché of pessimistic commentators who seldom attempt to clarify precisely what water-related causes of conflict are and what they are not; and how the link between physical processes of water scarcity and the rise in violent conflicts within or between societies should be interpreted. Indeed, the central claim of this book is that Middle Eastern water problems are not inherently different from those in other parts of the globe, and that the doom-laden hypotheses which represent the dominant view in the literature of hydropolitics are greatly exaggerated. We will argue that, far from leading to military conflict, increasing water scarcity will concentrate the minds of those involved to find sustainable solutions and, to achieve this goal, the concerned parties will increasingly resort to coordinated, cooperative, and conciliatory arrangements. Our claim has rarely been advanced in the literature. Deudney (1991: 26) has argued, though only in passing in a short article, that everyone's vulnerability to interruption of water supply will make war less likely More recently, Gleick 1994a, Elhance (1995), Dellapenna (1995), Wolf (1995, 1997) and Alam (1998) have explicitly questioned the thesis that severe water scarcity will lead to interstate conflict. In this book, following Deudney, Gleick, Elhance, Dellapenna, Wolf and Alam, we will demonstrate that water scarcity is more likely to generate cooperation than conflict.

Through an analysis of the evolving and dynamic process of hydropolitics in the major international water basins of the Middle

East, we will test our main hypotheses – that water is too vital a resource to be put at risk by war, and that water scarcity, as an *independent variable*, encourages cooperation between riparian parties of a shared water basin. The reasoning behind these hypotheses is that the aridity of the region, the acute needs of the region's nations, and the vulnerability of expensive water engineering systems such as dams, hydropower installations, and water distribution networks, create a mutual hostage situation in which all parties are at risk should there be any water war. Moreover, the increasing environmental awareness in the region has highlighted the fact that since war cannot change the ecological givens, it could not increase the water supply in real terms and in the long run the cost of war would far exceed the possible returns. Furthermore, the historical background of water management policies in the region indicates that the indigenous inhabitants of the Middle East have always been aware that cooperation between riparian parties over their shared water resources is the only way to create a win–win situation in which all parties are better off. The scarcity of water, therefore, though it might cause periodic tensions, does not encourage states of the region to employ violence to resolve the problem. Indeed, as we shall see in Chapters 4, 5 and 6, there is considerable evidence indicating that hydropolitics in the Middle East is a context for cooperation in which the development of common water resources will create a network of collective interests and a platform for common perceptions that will finally breed more regional integration and peaceful coexistence.

The framework of analysis

Since the nature of this study is multidisciplinary,[8] a multidisciplinary method of study is adopted. We take the view that there is no single approach or perspective or paradigm that can wholly explain this complex issue. Addressing the issue from a specific outlook can only help us to understand one aspect of its many dimensions. For instance, those who seek to explain the issue of water scarcity exclusively from a strategic point of view and in the context of power politics fail to understand other socio-economic variables which affect or are affected by the water issue and, hence, incline to gloomy predictions whose fallacy we disclose in later chapters.

Approaching the issue from a multidisciplinary perspective, we focus on the five main approaches to hydropolitics – namely the security, economic, legal, technological, and environmental approaches

– and argue that, while each approach offers considerable illumination, the fifth approach casts most light on the problem as it exists today. The security approach rightly warns of the dangers of escalation of water conflict, but too easily militarises the issue and exaggerates the danger. The economic approach rightly points to the need for treating water as a commodity but too readily overlooks insufficiencies of market forces in dealing with such a unique vital natural resource. The legal approach rightly argues that legal principles can be effective devices for settling competing uses of water, and truly exposes the paucity of international law on water rights, but ignores the fact that the first step in translating legal theory into practice is the formulation of political agreements through which the key institutional machinery for applying legal principles can be created. The technological approach rightly draws attention to supply-side inefficiencies, but addresses the symptoms rather than the causes. The environmental approach, while not denying the insights of the other approaches, penetratingly relates the water crisis to wider environmental issues, showing convincingly how the crisis has arisen and how it can be dealt with.

Each of these approaches is tested in relation to three case-studies of water politics in the Middle East – the Jordan River basin, the Euphrates–Tigris basin, and the Arabian Peninsula. These three cases have been chosen because they represent the most important cases of international water basins in the Middle East, whose dwindling water resources have been widely identified in the literature as the most explicit manifestation of the water crisis. Hence, they enable us to test the hypothesis of 'water war' in the Middle East. The cases are examined in order of the seriousness of water tension, beginning with the Jordan River basin, whose waters, owing to the severity of the Israeli–Palestinian conflict, have been heavily politicised, securitised, and militarised. In the case of the Euphrates–Tigris basin, despite relative abundance of water, because of political rivalry between riparian parties, water resource management has been turned into a matter of foreign policy and diplomatic confrontation, though the parties have been able to maintain an improving practical relationship. Finally, in the case of the Arabian Peninsula, which is one of the most acute water-deficit areas of the world, abundant petroleum resources have been used to avert a water crisis. Despite the severity of water scarcity in this basin, the concerned countries have always perceived the issue in an economic context rather than in political or security contexts. By examining these three different cases, our main hypothesis is confirmed – that water scarcity is more likely to lead to cooperation than to conflict.

Organisation of the chapters

In Chapter 2, we review the literature of hydropolitics and identify and discuss the five main approaches to the issue – the security, economic, legal, technological, and environmental approaches – and the conceptual paradigms from which they emerge. Linking the first and the last approach, two different frameworks of security thinking are identified – the traditional framework of strategic security thinking, and the new framework of environmental security thinking. It will be shown that most writers who paint a gloomy picture of the water conflict are looking at the issue from the traditional framework in which water is deemed a core national security problem. It will be argued, however, that such a perspective is unhelpful to an understanding of the issues involved. This is because no country in the region can solve its water problems independently; that is, without encroaching upon the resources of its neighbours. Therefore, the shortage of water, if perceived as a national security issue, is bound to exacerbate further tensions and conflict. This leads the regional countries into a vicious and self-destructive circle; no comprehensive water development, which is for the benefit of all, can take place without peace, and, conversely, no peace is possible or sustainable without comprehensive water development.

Looking at the issue from the framework of environmental security, however, turns the very problem that engenders rivalry and threatens to instigate war, into a powerful inducement to promote peace. Instead of seeing water security as a *national* security issue (state versus state), the environmental approach sees it as a *global* security issue (humanity and nature), and as a result, rejects military solutions, since the resort to war merely exacerbates the global security problem. The only sustainable solution to water scarcity is for states to cooperate together through regional and international agreements. We seek to prove to the sceptics, who pride themselves on being 'realistic', that this environmental approach is not utopian. Indeed, on this view, water cooperation will prevail in the Middle East, because the only alternative – a perpetuation or escalation of conflict into open warfare – is self-evidently unrealistic.

A detailed and comprehensive inventory of the water scarcity issues throughout the world is beyond the scope of this study. However, some important cases outside the Middle East are presented in Chapter 3. After conceptualising the terms 'water scarcity' and 'global problem', the water resources management practices in each region are examined, focusing particularly on Europe, North America, China, the Indian subcontinent, and Central Asia, before turning to the Middle East. The chapter concludes that, although there is every reason to believe that a looming water crisis is confronting humanity, the challenge is more regional than global in scope.

Chapters 4, 5, and 6 analyse the water politics in the major international water basins of the Middle East covered by the Jordan River and its surrounding areas; the Euphrates–Tigris Basin; and the Arabian Peninsula. In these chapters the general climatic and hydrological conditions of each basin are briefly noted; the impact of hydrological givens on the sociocultural, economic, and political developments of the riparian communities of each basin is evaluated; their old and new water management policies are examined; the effect of these policies on their foreign policy considerations is highlighted, and the reasons behind the lack of comprehensive agreement among the riparians of each basin are identified.

In the concluding chapter, the significance of this study and its endeavour to put forward a fresh interpretation of water politics is evaluated. It will establish that, although water has sometimes provoked dispute in the Middle East, it has much more often promoted coexistence and cooperation between indigenous communities. The main hypothesis is that, even though the most frequently mentioned scenario in the literature is that disputes over water supplies will lead to interstate conflict and war, on the contrary, it is most unlikely that the quest for water will cause a new war in the Middle East. Indeed, water scarcity should be seen as a platform for regional cooperation that promises optimal development and exploitation of the region's water supplies through which all riparian nations can achieve greater gains. As a win–win strategy, the joint cooperative development of common water resources will reinforce peace, not provoke war.

Notes

1 The *Code of Hammurabi* is the most nearly complete collection of Babylonian law yet discovered and one of the most important ancient codes of law. Hammurabi's code of law consisted of 282 provisions systematically arranged under such headings as family, labour, personal property, real estate, trade, and business. This code is believed to have greatly influenced the development of Near Eastern civilisation for centuries after it was written. A tablet bearing the code was discovered in Susa, Iran, in 1901.

14 *Water Politics in the Middle East*

2 For a brief review on the Islamic legal tradition connected with water rights and water management, written in English, see Caponera (1973).
3 See, for example, Abramovitz 1996. Agnew and Anderson 1992; Berkoff 1994; Bhatia *et al.* 1993; Biswas 1978, 1991b; Clarke 1991; Davey 1985; Dinar and Loehman 1995; Falkenmark 1984, 1986b, 1990; FitzGibbon 1990; Gleick 1993b; Kirmani and Rangeley 1994; Le Moigne 1994a, b; Meybeck *et al.* 1989; Munasinghe 1990; Postel 1989, 1992; Pringle 1982; Rogers and Lydon 1994; Science Council of Canada 1988; Serageldin 1994, 1995; Serageldin and Steer 1994; Shahin 1989; Stumm 1986; Thomas and Howlett 1993; Thompson 1978; Winpenny 1992, 1994; World Bank 1990, 1992b, 1993.
4 See for example: Allan with Court 1996; Allan and Mallat 1995; Anderson 1991a, b, c, d, 1995; Biswas 1994; Bulloch 1993; Bulloch and Darwish 1993; Charnock 1984; Cooley 1984; Downey and Mitchel 1993; *The Economist* 1987; Falkenmark 1986a, b, 1989; Frey and Naff 1985; Gischler 1979; Gleick 1994a, b; Gowers and Walker 1989; Guttman 1996; Hatami and Gleick 1993; Hillel 1994; Irani 1991; George Joffe 1993; Kliot 1994; Kolars 1991; Lean 1993; Lonergan 1997; Lowi 1984, 1993a, b, 1995; Mooradian 1992; Pearce 1991; Pearce and Hudson 1991, Perera 1981; Pope 1995; Richards 1992; Roberts 1991; Rogers and Lydon 1994; Schmida 1984; Shadid 1995; Shapland 1997; Starr 1991, 1993, 1995; Starr and Stoll 1988; Thompson 1978; World Bank 1994, 1995; World Water 1984.
5 For instance, *Water International*, 1993, **18** (1), *Environment*, 1994, **36** (3), *Research and Exploration*, 1993, **9** (special issue), and MEED 1995.
6 See also Ben-Yamin 1991; Falkenmark 1989; Frey and Naff 1985; Lowi 1993a, b; Thomas and Howlett 1993; Wolf and Dinar 1994.
7 Anderson 1991a, b, c, d; Cooley 1984; *The Economist* 1987, 1990; Elliott 1991; Gleick 1994a; Gowers and Walker 1989; Pearce 1991; Pearce and Hudson 1991; Pringle 1982; Richards 1992; Thompson 1978.
8 Consider its links to various disciplines including agriculture, industry, energy, political economy, health and settlement of population, social development, security, ideology, and the environment.

2
Five Approaches to Water Scarcity

Introduction

Shrinking and deteriorating fresh water reserves, increased flooding and droughts, and rising water budgets that are hardly affordable – all are clear signals of a water crisis facing humanity in different parts of the globe, embracing both developed and developing countries. These facts suggest that the growing problems in the water domain need to be better understood if they are to be solved. The literature on this issue reflects a realisation of this need. The UN has long been stimulating research into the water crisis, and, according to Falkenmark's assessment (1990: 188), 20 UN bodies are now involved with water issues. In addition, many academic institutions, centres, corporations, consulting firms, foundations, national and international agencies, non-governmental organisations (NGOs), non-profit and private organisations have also recently become interested in hydropolitics and water-related issues. As a result, there are scores of published studies, as well as a burgeoning literature on water issues available on the Internet.[1]

In this chapter we identify the five main approaches to the issue of water scarcity in the literature on hydropolitics – security, economic, legal, technological, and environmental. First, there are writers who argue that water is a source of power and that water scarcity is a critical and highly strategic issue which affects the social and economic development of nations and, consequently, threatens to undermine their political power. Looking at the issue in the context of a realist view of politics, these writers confidently predict that states are prepared to go to war to ensure a steady supply of water for their present needs as well as for their projected demands. Second, there are economists, who, rejecting the alleged strategic importance of water as a unique resource, argue that water scarcity is basically an economic problem which will be alleviated if nations treat water as an economic asset. Thus the solution lies in the application of the market mechanism, not in the use of military force. Third, there are writers who approach the issue from a legal perspective, arguing that the success of the market mechanism depends on the existence of proper institutions which provide an enabling environment for the implementation of economic rules or any other management mechanism. One of the most important of these institutions is a settled and clearly defined system of property rights. From this perspective, the absence of proper international agreements among riparian nations concerning their water rights is the root cause of the water crisis, and its successful resolution rests in the establishment of water rights at the national level as well as the international level. Fourth, there are the technological optimists, who, denying the whole idea of resource scarcity, argue that with the help of technology there is more than enough water available to meet all the conceivable needs of present and future generations. They claim that a *technological management* of water resources with the help of the market mechanism can surely solve the problem of water scarcity. Finally, there are the environmentalists, who believe that the water crisis is part and parcel of the global environmental crisis that we are facing today. Rejecting the validity of the traditional notion of security, the competence of market mechanism *per se*, the sufficiency of legal frameworks, and the adequacy of technological solutions, environmentalists introduce the notions of *limits to growth, sustainable development*, and *environmental security*. They argue that the Earth is a finite planet with limited resources. This finitude, and the scarcity it implies, places limits on our use of its resources, beyond which the Earth fails to recover and restore its equilibrium. From this perspective, water scarcity is an environmental problem which stems from our misguided conduct toward nature and our unsustainable policies for exploitation of this delicate resource. The resort to military, economic, legal, or technological solutions not only cannot solve the problem but actually exacerbates the predicament. The solution is to understand the limits to growth of the eco-geographical region in which we live and to create a sustainable society accordingly.

Detailed discussions of these five approaches in this chapter leads to the conclusion that, since water plays a multifunctional role in shaping the culture, religion, economy, nutrition, health, and every other aspect of human life, water-related issues need to be seen in the context of the relationship between humans and their 'living environment' as a whole rather than in isolation as a military, economic, legal, or technological problem. This implies that water scarcity, as an environmental issue, should be looked at from the perspective of 'environmental politics', which is very different from the traditional perspective of 'power politics'. Let us explain this important point before examining in detail each of the five approaches in turn.

Since environmental politics is committed to an interdependent and global conception of environmental security as an alternative to the prevailing concept of national security (Hugh Dyer 1996: 23), and strongly endorses collective solutions, it serves to shift policy-makers away from conventional realist positions based on protecting their narrowly defined national interests by the application of strong military capabilities. Viewing the issue of water scarcity from the perspective of environmental politics turns the predicament into an opportunity for cooperation and a powerful inducement to promote peace. Those who look at the issue from the traditional perspective of state-centric security discourse with its emphasis on state interests and the accretion of state power in an anarchic system, suggest that water scarcity is a core national security issue which generates conflict and instigates war. However, there is a lack of hard evidence to substantiate this realist view. Examples are quoted of competition and contentiousness over shared water resources, and cases in which shared water supplies have been used as instruments of politics or as weapons in conflicts generated by factors other than water scarcity, or in which military attacks on water installations and water distribution facilities have occurred (see Gleick 1993a: 87, 1994a: 15). These examples are cited to illustrate the causal linkage between water scarcity and conflict, but very seldom is any attempt made to clarify precisely what water-related causes of conflict are and what they are not; and what the causal link between physical processes of water scarcity and the rise in violent conflicts between countries actually is. Because of the lack of analytical precision, a point to which we shall return later in this chapter, the concept of *water conflict* is ambiguously used in the literature. If we look closely at the examples of so-called water conflicts which are cited by writers who look at the issue from a narrow, militarised, security perspective, we usually find extremely complex situations in which the relationship between cause and effect is not as clear as it has been assumed. Jumping to hasty conclusions has led these writers to oversimplification and false hypotheses, as we shall see. Moreover, an examination of the proposed solutions for the management of water crisis indicates clearly that there are several feasible choices and opportunities for policy-makers by which they can ease their water scarcity problems without becoming involved in violence and hostility. Furthermore, the experience of the Middle Eastern countries during the last three decades, which is examined in chapters 4, 5, and 6, buttresses our central hypothesis that water scarcity encourages coordination and, as an *independent variable*, will lead to cooperation between countries over optimal management of their water resources. This is the only option which is to the benefit of all riparian states in an international water basin and is the ideal solution for reconciling the competing needs and conflicting interests of all parties.

Five hypotheses about the water issue

Water as a security issue

A realist view of world politics suggests a positive relationship between resource scarcity and conflict. From this perspective, human history is an account of resource wars (see Westing 1986). Applying this analysis to fresh water resources, Falkenmark and Widstrand (1992: 4) argue that 'access to water has generated political and military conflicts throughout world history'. Military analysts such as Thompson (1978: 62–71) claim that 'fresh water is similar in many respects to other of the world's scarce resources' and since 'fresh water is becoming increasingly scarce', it is thus becoming 'increasingly a source for future conflict'. In the early 1980s, a series of research studies on military activities and the human environment carried out jointly by the Stockholm International Peace Research Institute (SIPRI) and the United Nations Environment Programme (UNEP), concluded that the geographical distribution, availability, and degradation of the world's natural resources, among which oil and minerals, fresh waters, ocean fisheries, and food crops were singled out for detailed analysis, 'influence the international security perceptions that govern strategic policies' of nation-states and dictate 'the use of military force' by them (Westing 1986: v).

Falkenmark sees the fresh water issue as a present and future focus of international disputes and as a factor in conflict formation.

Five Approaches to Water Scarcity 19

She argues that frustrations over scarcity of water and over dependence for water upon upstream countries may develop into disputes. 'Water can be a strong contributing factor to armed conflict,' she believes, 'even if this is not often recognized' (1986a: 109). From a similar perspective, Gleick (1993a: 79) argues that, as the twenty-first century approaches, water and water-supply systems are increasingly likely to be objectives of military action, instruments of war, and a salient element of interstate politics. Emphasising the role of resource security as a crucial factor which affects both national and international security, Gleick asserts that 'the focus of security analysts must now be *when* and *where* resource-related conflicts are most likely to arise, not *whether* environmental concerns can contribute to instability and conflict'. Underlining water as a resource which fits into the traditional framework of security analyses, he claims that 'if water provides a source of economic or political strength', which obviously it does, 'ensuring access to water provides a justification for going to war and water-supply systems can become a goal of military conquest' (Ibid.: 82–4).

According to this perspective, which is the dominant view in the literature, water scarcity is a very critical issue that affects the national security and survival of each society. To substantiate this argument, it has frequently been claimed that water disputes were the principal cause, or at least one of the main causes, of the 1967 Arab–Israeli War.[2] A systematic examination of the 1967 war, however, reveals that the water issue was only a *contributory* factor in that conflict, not a *causal* determinant of it (see Chapter 4). Moreover, these writers typically overlook the fact that many countries in which water is the most critical resource constraint, quite often cooperate actively in increasing their shared water resources. For example, Egypt and Sudan have long cooperated on the Nile River, based on the Aswan Dam and Lake Nasser, and on the Jonglei Canal project. Their agreement not only governs the sharing of the Nile's waters, but also contains an instrument for settling controversies by negotiation (Howell and Allan 1994). Mali, Mauritania, and Senegal have cooperated for nearly a quarter of a century in building dams and managing the Senegal River through a partnership scheme; and the eight countries of the Zambesi Basin have tried to conclude an equitable and environmentally sound river development plan under the auspices of the United Nations Environment Programme (Postel 1992: 86; Lean 1993: 21). Four case-studies by Cano (1986), Chomchai (1986), LeMarquand (1986) and Mehta (1986) demonstrate that established agreements on transboundary water resources in Third World countries exhibit a general spirit of non-confrontational approaches to water issues and a willingness to accommodate variations within a carefully designed system of rules and regulations. In fact, there is even a history of tactical cooperation over water, albeit limited in scope, between antagonistic neighbours such as India and Pakistan (see Alam 1998; Mehta 1986; Kirmani 1990; Biswas 1992), Iraq and Syria (see Chapter 5), and Israel and Jordan (see Chapter 4).

As the above examples indicate, not all water-resources disputes lead to violent conflict; indeed most lead to discussions, negotiations, mediations, and non-violent resolutions. Moreover, given the general atmosphere of tension in the Middle East, it is remarkable that so little open conflict over water has erupted in this volatile region. Paradoxically, complexities and tensions raised by hydrological problems have often tended to compel cooperation where other non-water antagonisms have degenerated into warfare (Naff and Matson 1984: 4). Therefore, it can be argued that there is no iron law determining that water scarcity *per se* must inevitably lead to destruction, competition, or violent conflict. This argument has been briefly advanced by Joseph Dellapenna, Arun Elhance and Daniel Deudney. For instance, Dellapenna (1995: 89) argues that 'the very importance of water makes co-operation over water more likely than conflict'. He maintains that 'as in ancient times, the shared need for optimum management of this scarce resource can become a source of regional unity rather than regional discord'. Similarly, Deudney (1991: 26) contests the idea of 'water war' arguing that, 'It seems less likely that conflicts over water will lead to interstate war than that the development of jointly owned water resources will reinforce peace.' He explains that:

> Exploitation of water resources typically requires expensive – and vulnerable – civil engineering systems such as dams and pipelines. Large dams, like nuclear power plants, are potential weapons in the hands of an enemy. This creates a mutual hostage situation which greatly reduces the incentives for states to employ violence to resolve conflicts. Furthermore, there is evidence that the development of water resources by antagonistic neighbors creates a network of common interests. (Ibid.)

Looking at the issue from a realist perspective, the problems of water scarcity are most acute in the context of the upstream–

Five Approaches to Water Scarcity 21

downstream relationship which is perceived as a power relationship. In this context, 'as demand for water grows, and supplies shrink, political power swims upstream' (Lean 1993: 19). Thus, water security, which is viewed as part and parcel of national security, is sought through zero-sum thinking and a 'we versus they' mentality. Falkenmark and Widstrand (1992: 5), for instance, see 'the history of water and populations' as 'a history of power in society'. Approaching the problem of water scarcity in this context, Falkenmark argues that since many river systems and large aquifers are shared by several countries, this geographical fact 'adds to the risk of international disputes or even confrontations, especially in view of sharply divergent interests in the shared resources of the upstream and downstream countries' (Falkenmark 1990: 177). Accordingly, the highest levels of frustration that will arise from increasing water scarcities can be expected to develop in countries where populations are increasing dramatically and where most of the water sources are transnational (see Falkenmark 1986a: 109 and Gleick 1993c: 108–10).

The analogy with oil

To explain the connection between water security, as a subset within the overall field of national security, and international violence, many realists draw an analogy between water and oil. For example, McDonald and Kay (1988: 2) predicted that 'the energy crisis of the 1970s will take a back seat to the water crisis of the 1980s and 1990s'. Similarly, Biswas (1991b: 144) claimed that 'by the end of this decade, water will be *the* critical resource problem of the world, somewhat similar to what the energy crisis was in the 1970s'. The Gulf War, which appeared to exemplify the notion of 'resource war', has been used by many scholars as an analogue for forthcoming water wars. For instance, Sandra Postel of the Worldwatch Institute declares that, 'Water has become a strategic resource like oil, for which nations will compete fiercely as it becomes more scarce. And like oil, it is likely to lead to warfare as oil did in 1991', and Linda Starke also supports this hypothesis (see Postel 1992: 13 and 188). Butler (1995: 34) goes further, arguing that 'security of water sources is already one of the most crucial elements in foreign policy for many countries, far out-stripping the importance of oil'. 'After all', he notes, 'alternative energies can always be found – be they gas, coal or nuclear – but without water, life itself is impossible.' Likewise, as we noted in Chapter 1, Ismail Serageldin, the vice-president of

22 Water Politics in the Middle East

the World Bank, issued a prophetic statement that: 'Many of the wars this century were about oil, but wars of the next century will be over water' (*Financial Times*, 7 August 1995, p. 2).

The analogy with oil, however, is not entirely helpful to the realist argument. As Lipschutz and Holdren (1990: 125) point out, the oil experience should have taught us that, when the economic or political costs of the usual sources of supply rise too sharply, other sources previously considered uneconomic may come into play. In the case of water, the alternatives include not only lower-quality water resources, such as brackish water, that are more widely distributed than high-quality fresh water, but also *recycling*, *substitution*, and *conservation*, which could help in optimising applications to obtain increased functional benefits from each unit of the available resources. Moreover, a dramatic reduction in demand can be effected by changes in both technology and custom. According to Clarke (1991: 71),

> contrary to popular belief, human attitudes can change very swiftly in the face of crisis. Who would ever have guessed, for example, that the energy-dependent West could have made such dramatic reductions in energy use during the oil crisis of the 1970s? The same is certainly possible with water.

An interesting study by Mori (1987) strongly supports this argument. In his examination of Japanese problems concerning water quantity, from the standpoint of supply and demand, Mori observed that, following the oil crises of the 1970s, attitudes to water conservation changed. Some industries reduced their consumption by more than 80 per cent, and domestic consumption fell in certain cases by more than 30 per cent.

Moreover, a review of the recent literature on environmental change and conflict (see below) reveals that the realist perspective is unable to explain the ramifications of environmental crises in the international arena. This is because from a realist perspective states are preoccupied with the expansion of their power in an anarchic system in which their performance is mainly a function of the structure of power relations. Such an emphasis on 'state' and 'power' persuades realists to view countries not as interdependent ecological units, a central notion of environmental politics, but as exclusively divided territories governed by autonomous states. But given the transboundary nature of most environmental problems, realism cannot confine them by its traditional concepts – 'state', 'sovereignty', 'territory', narrowly defined 'national interest', and 'balance of power'. Realists fail to see that significant aspects of environmental problems – their transboundary nature and their interconnectedness – undermine the ability of states to solve them independently or by the application of military force.

We shall return to these environmental issues when we discuss the fifth hypothesis. In the meantime we must consider another critique of the realist perspective – that advanced by economists who argue that conflict over scarce water resources can be averted by giving water a monetary value.

Water as an economic issue

'Water, water everywhere but at a price.' This is the slogan of those who look at the issue from an economic perspective. The economics of water is being increasingly emphasised as a basis for the peaceful regional management of water resources. In the economic sense, water is a commodity just like other goods that we need to satisfy our requirements (Hirshleifer *et al.* 1969: 2). The core of this argument is that in many cases the failure to treat water as a commodity lies at the heart of the problem. For the economists, the problem of water scarcity could easily be solved, since water is a renewable and reusable resource, and worldwide there is more than enough water available, provided that people are willing to pay the costs of reclamation (see Table 2.1). From this perspective, turning water into a commodity will ease water stress; reduced water stress, in turn, will alleviate user conflicts at local, national and international levels (Winpenny 1994: 110–11).

To establish this hypothesis, some economists have challenged the notion of the *sensitivity* or *unique importance* of water which is postulated by those who regard it as a security issue. Hirshleifer *et al.* (1969: 4) argue that merely because a nation or an individual could not survive without water does not show its uniqueness. There are other commodities, such as energy and clean air, that are equally essential to human existence. While they do not deny the special features of water – including its open access, natural origin and availability without cost in some locations but expense to transport to others – Hirshleifer *et al.* (Ibid.: 5) contend 'whatever reason we cite, however, the alleged *unique importance* of water disappears upon analysis.'

From the economic perspective, as pressure of demand upon water supplies increases, the price of water should rise. Those who can

Table 2.1 Global water volumes

	Volume (km^3)	%
Oceans	1 348 000 000	97.39
Polar icecaps, icebergs, glaciers	27 820 000	2.01
Groundwater, soil moisture	8 062 000	0.58
Lakes and rivers	225 000	0.02
Atmosphere	13 000	0.001
Total	1 384 120 000	100
Freshwater	36 020 000	2.6

Freshwater as % of its total	%
Polar icecaps, icebergs, glaciers	77.23
Groundwater to 800m depth	9.86
Groundwater from 800m to 4 000m depth	12.35
Soil moisture	0.17
Lakes (fresh water)	0.35
Rivers	0.003
Hydrated earth minerals	0.001
Plants, animals, humans	0.003
Atmosphere	0.04
Total	100

Source: Baumgartner and Reichel (1975: 14)

take water for higher value uses will be ready to pay for it. Those who cannot afford the higher costs would give up their supplies to those who can. The distribution depends on the marginal value of water in each society. Therefore, 'even should the amount of water available to society as a whole be fixed, the amount available to any given use or user could be increased – at a price' (Ibid.: 2–3). This is true at international as well as at national level. For example, Americans use the waters of the Colorado River so heavily that the water becomes very salty when crossing the border into Mexico. However, they have been ready to finance a huge desalination plant to reduce the salinity of water at the border. Long-distance water transfer schemes in countries of Central Asia in the former Soviet Union, the great Man-Made River in Libya, the National Water Carrier in Israel, the GAP project in Turkey, and the Third River project in Iraq, are all based on such marginal value difference. Even if, for political or technical reasons, water transfer *per se* is not feasible, 'water shortages are not determining constraints because there is a ready supply of water, albeit "virtual water", available. This water is contained in the imports of staple foods which are for the moment readily available and massively subsidised on the international market' (Allan 1994c; cf. 1995). This argument is well substantiated in the current circumstances in the Middle East. Despite the implicit tension over water in the Middle East, which many predicted would be the region's most politically divisive and destabilising issue during the 1990s, the region has not faced any violence over water since 1970 (see Chapters 4, 5 and 6). This is partly because, according to Allan and Mallat (1995: 2), water proved to be vicariously available in the world market for food staples; and partly because the decision-makers in the region learned in a very practical way to substitute for water. They have begun to adjust their water allocation and management policies to achieve effective use of their water resources, which are fortunately more than adequate for their domestic and industrial needs.

Among several economic interpretations of the water resources issue, a recent study by Winpenny (1994) is very persuasive. He claims that for a long time water has been wrongly treated as if it were plentiful and free. The failure of both suppliers and consumers to regard water as a 'scarce commodity' and to recognise its 'economic value' lies at the heart of the crisis. His analysis suggests an emphasis on a demand management approach, as opposed to a supply augmentation approach, and on the greater use of water prices and market incentives to change the way people view and use water. Winpenny sees a link between failure to recognise the economic value of water, and wasteful and environmentally damaging uses of the resource. He argues that water authorities have been always obliged to supply consumers' needs, with cost a secondary consideration. Subsidised by their governments, consumers have been accustomed to treat water as an automatic public service that is supplied at zero or low cost to them. As a result, he believes, 'water is becoming one of the largest, and certainly the most universal, of problems facing mankind as the earth moves into the twenty-first century'. Although the problem varies in nature and intensity, he points out that 'it is challenging governments of countries at all stages of development, in most parts of the world' (Winpenny 1994: 1). Let us examine in more detail Winpenny's work as a paradigmatic economic analysis of the water issue.

Economic analysis of the causes of water scarcity

From an economic perspective, there are three causes underlying the problem of water scarcity. First, water is underpriced compared to its real cost of provision; second, water is underpriced compared to its environmental costs; and third, water is often a public good which makes it difficult to extract an economic price from users. Underpricing of water as an economic resource is the most basic reason why inappropriate habits of supplying and using water have persisted. In elementary economics, a commodity which is supplied too cheaply will eventually become scarce, and will need to be rationed by more or less arbitrary means. Those who are fortunate to be provided with it will use the commodity excessively and wastefully at the expense of others who do not have access to it. This is just the case in the water domain. Users do not treat water as an economic (scarce) commodity because it is cheap and ready at hand, while in most situations water has a significant real cost of supply. Winpenny (1994: 9) claims that the market could be used as a means of solving the problem of scarcity by including the real costs of water supply in its price, including the 'provision' costs, 'opportunity' costs, and 'environmental' costs. The cost of providing comprehensive public water services includes capital and recurrent outlays in supply, treatment and distribution, as well as drainage, sewage collection and treatment, flood control, and recreation (for a detailed explanation of the principles involved in deriving prices from these costs see Warford 1968). To optimise the use of water, tariffs should be based on the marginal cost of supply: that is, the cost of adjusting long-term capacity caused by a given change in demand. The argument is that the benefit from the marginal unit of supply should equal the cost of providing that unit. If the benefit were less, society would gain by reducing supply to the point of equality. If the benefit were greater, there would be gains from expanding supply (Winpenny 1994: 9). Other writers – such as Repetto (1986), Postel (1989), and Katko (1990) – indicate that, in every developing country and most developed ones, prices for urban and industrial users usually fall well short of the economic costs of provision, and the situation with reference to agricultural users is even worse.

Opportunity costs mean sacrificing the value of water in alternative uses. If, like other goods, water were traded according to demand, these alternative values would be reflected in its price. It is the lack of a water market system that prevents its opportunity cost getting across to the consumer or the polluter. Although Winpenny (1994: 10) admits that, since water's opportunity cost varies sharply between different users and at different times, it is impractical to incorporate it into a standard pricing formula, he insists that 'the

Five Approaches to Water Scarcity 27

existence of alternative uses for water, which are often more valuable than that of the target consumers, reinforces the case for charging at least the economic price'. He maintains that the gap between the opportunity cost and the actual cost of water is currently so wide that any change which brings prices into the right general range is justifiable.

Finally, environmental costs of water have been basically neglected, even though water consumption in every sector incurs its own environmental consequences. For instance, aquifers which are drawn down at a rate faster than their natural recharges, are exhausted or contaminated to the point of catastrophe. Agricultural run-off returns to rivers with high saline and chemical content, which is costly to households, other farmers, and fishers. Untreated industrial effluent can also poison fish, spoil rivers and lakes, and impose heavy treatment costs on society. Diversions of water for power generation or irrigation reduce river flows, and can cause siltation, transport difficulties, desiccation of wetlands, loss of fishing grounds, and the destruction of amenity. Several writers have tried to measure and subject these effects to economic valuation (see, for example: Kneese 1984; Gibbons 1986; Winpenny 1991).

However, even if society wanted to recover the provision, opportunity and environmental costs of supply and to charge polluters, the 'public-good' nature of water often makes this difficult. The essence of a public good is that it is available to all, and no one can be denied access to it (Ostrom 1990). Thus as the 'tragedy of the commons highlights, no single agent has an incentive to refrain from exploiting it, since others would continue to do so. Moreover, a private agent has no incentive to invest in its preservation, because it would be impossible to recover costs from free-riders.

Winpenny realises that certain water resources are public goods in this sense – for example, an aquifer underlying a number of territories, or unmetered piped water in the mains. The public-good nature of aquifers needs not to be spelt out. Unmetered piped water is also a public good, in the specific sense that the decision to consume more is costless to the individual user and the water authority has no sanction on excessive and wasteful consumer(s). A metering system, however, coupled with a proper charging structure, could solve the problem in the case of piped water. By using water meters, Davey (1985: 16) notes, the Romans used the market mechanism to moderate water usage. Regarding aquifers, theoretically, 'it is possible to control extraction by co-operative agreement, legislation governing sinking new wells, regulating extraction, adjusting the energy prices for pumping, etc.' (Winpenny 1994: 12). Practically, however, this is difficult and the widespread depletion of aquifers in different parts of the world is evidence of this difficulty. Nevertheless, Winpenny asserts that 'water is not bound to remain a public good', and it is possible to bring it under greater control for the public benefit. This, of course, requires acceptance by public opinion; passing of relevant legislation; and installation of administrative, policing, and charging structures far beyond those now existing in most countries.

Inadequacy of supply-side remedies for water scarcity
Winpenny exposes the inadequacy of the traditional approach to water scarcity – the supply-side solution. He argues that the supply augmentation approach is unsustainable in many countries for three main reasons – hydrological, environmental, and financial limitations. Regarding hydrological limits, as Chapter 3 illustrates, in most parts of the world the readily accessible water sources are approaching their physical limits and new supplies are available only at rapidly increasing cost. As for environmental costs, measured in economic terms, Winpenny claims that the environmental costs of new water supply schemes are becoming less acceptable. These costs arise both on the supply side – depleting aquifers, damming rivers, destroying wetlands – and on the disposal side – runoff, effluent and sewage control and treatment (Winpenny 1991). Taking the desiccation of the Aral Sea in Central Asia as an example of environmental catastrophe, he argues that if the true environmental costs of long distance water transfer as practised in this region had been better understood at the time, a different course of development would surely have been followed.

With regard to financial constraints, Winpenny (1994: 5) declares that 'water utilities, and their governmental sponsors, are in no position to bear the increasing capital, operating and maintenance costs of catering for the projected growth of water requirements'. Supporting his argument from a review by the World Bank, in which the effective price per unit of water was found, on average, to be only one-third of the full economic cost of supplying that unit, he attributes the poor financial performance of water utilities to underpricing and weak cost-recovery systems, and to the high proportion of leaks and wastage. Also Winpenny points out that the public sector can no longer bear the sole burden of investment for water requirements in industrialised societies, let alone in developing countries.

The low price charged to consumers discourages the technical changes in water-user sectors that are necessary for the sake of long-term conservation of this increasingly scarce resource. Moreover, conservation by users – whether farmers, firms, or domestic users – is not worth while so long as prices remain low. In agriculture, which accounts for two-thirds of the world-wide use of water (Postel 1992: 99; Gleick 1993b: 56), there is enormous waste in its distribution and application. Most of the water is lost to leakage or evaporation before it gets to the field. According to Postel (Ibid.: 100), world-wide irrigation efficiency is estimated to average less than 40 per cent. Even in the regions where water is actually very scarce, over-expansion of subsidised irrigated farming consumes the lion's share of the available water resources. There is a professional consensus now that making irrigation more efficient is a top priority in moving toward sustainable water use. If a proper tariff system encourages farmers to reduce their water needs by only 10 per cent, for example, it would provide up to double the domestic water use world-wide (Ibid.: 99).

In industry, also, the extent of waste and misuse of water is evident in the large savings that are possible, some at little cost. Winpenny refers to examples of major savings being achieved through industrial recycling which were only worth while to the firms once an economic tariff for fresh water was fixed, or where pollution charges or other controls were introduced. Major water-using and polluting industries, however, have been indulged through policies of protection and import substitution; the price of their water and pollution has not been high enough to influence their viability or growth.

In the domestic sector, underpricing urban household water is said to have encouraged the over-expansion of cities. Conurbations such as Mexico City, Santiago, Beijing, and Delhi are starting to experience major water shortages, caused in part from the 'subsidy' to their expansion, due to the failure to charge residents and commercial users the full cost of their water (Winpenny 1994: 17). Besides, experience shows that when a scarce commodity is not allocated by the market mechanism, it tends to gravitate towards the rich and powerful. Winpenny points out that in urban areas the gross under-supply of poor neighbourhoods coexists with high per capita levels of consumption among those with individual connections. He also points out that in high-income areas within arid or semi-arid regions a large part of peak-time consumption is for 'frivolous' purposes such as watering lawns and swimming pools.

In short, underpricing water accelerates the rate at which both available surface and renewable groundwater sources are allocated, and often leads to an irretrievable change in the hydrological system of watersheds and mining of aquifers. It also, as mentioned above, provides benefits for the rich and powerful and, consequently, leads to 'politicisation' of water allocation procedures. Politicisation, in turn, leads to increasing diversion of public investment resources into supply-augmentation projects, and a high proportion of the national budget devoted to subsidising the user sectors. That is a major economic distortion in many countries. For example, in Israel the agriculture sector has traditionally had a very powerful lobby in policy-making, and this is why still more than 79 per cent of its water resources are heavily subsidised and allocated to agriculture (World Bank 1995: 7). In Syria, according to Bulloch and Darwish, since the Euphrates valley is a major recruiting ground for the armed forces, priority has been given to irrigation in that area. However, they believe, 'it might be better used elsewhere'. Similarly, in Egypt 'veterans of the 1973 Arab–Israeli War were given plots of reclaimed land designated for increased food crop cultivation despite their lack of farming expertise; it was more important to remove potential dissidents than to ensure high production' (Bulloch and Darwish 1993: 188).

The demand management approach

These are powerful arguments for adopting an alternative approach to water management – that of demand management. This entails influencing existing water consumption patterns by taking into account the value of water in relation to its real costs and introducing measures which require consumers to relate their usage more closely to those costs. This approach was endorsed by the United Nations Conference on Environment and Development (UNCED) in the following terms:

> Pursuant to the recognition of water as a social and economic good, the various available options for charging water users (including domestic, urban, industrial and agricultural water-user groups) have to be further evaluated... A prerequisite for the

Five Approaches to Water Scarcity 31

sustainable management of water as a scarce vulnerable resource is the obligation to acknowledge in all planning and development its full costs. Planning considerations should reflect benefits investment, environmental protection and operation costs, as well as the opportunity costs reflecting the most valuable alternative use of water ... The role of water as a social, economic and life-sustaining good should be reflected in demand management mechanisms ... and implemented through ... financial instruments. (UNCED 1992, Agenda 21, Chapter 18, para. 15–17)

Such measures, it is claimed, would not only enable us to release relatively large amounts of water for redeployment elsewhere, but would also lead to a more equitable distribution of the resources. This notion of equity is one that a purely market-based water distribution system would not, however, achieve. A system of unrestricted economic pricing might solve some problems of water shortage, but it might also lead to the rich monopolising water and the poor becoming more likely to resort to violence. In order to prevent this, a combination of a life-line, low, flat rate and progressively higher rate charges for water consumption would be necessary. Since in most existing systems available resources are disproportionately used by the more affluent consumers, reducing subsidies and restructuring tariff systems according to a cheap flat rate and a progressive rate would, first, release funds for the expansion of the supply network to the poor, and second, work as a shield for poorer, or smaller, users from hardship. This system would encourage minimum levels of consumption, discourage wasteful and anti-social uses, release water for new users, and help to distribute available supplies more evenly and equitably.

What economists call an 'adjusted market' approach to water resources management therefore seems the most promising approach. However, markets and prices are only effective if they are used in the context of an enabling environment. Conservation measures, therefore, entail a package of economic and non-economic instruments. In other words, the utility of a market and pricing system depends on change in the way we view and use water. Radical reforms in attitudes, policies, institutions, and technology affecting the water sector are prerequisites for the creation of a macroeconomic context in which prices can function effectively. Market forces will be frustrated unless these other elements are present in the policy framework. Market incentives need especially to be reinforced by legal measures. This brings us to the third hypothesis – the legal perspective.

Water as a legal issue

From the economic perspective, then, water scarcity is really an allocation problem, not a supply problem. To solve the allocation problem, economists prescribe a set of economic measures, including market incentives and pricing policies. However, they have to admit that an 'enabling environment' is a prerequisite for the applicability of these economic measures (Winpenny 1994: ch. 3). Among the necessary elements which contribute to creating such an environment, appropriate legislation for establishing 'water rights' – both at national and international levels – is most essential. A legal framework is one of the basic prerequisites for the functioning of a water market because, as Lundqvist (1996: 13) points out, without having proper rules about who enjoys water rights and what the rights entail, trading is hardly possible. This fact and other policy issues dealing with the notion of 'water rights' are highlighted by those who look at the water scarcity issue from a legal perspective.

From this perspective, the common property nature of most water resources is a significant challenging factor. This is because the development of water resources to maximise their economic returns requires certainty of ownership and flexibility in their use. Without these two conditions there is little incentive for those who have access to water to conserve it for future use, invest in water reclamation, or exchange it for a higher valued use. Existing systems of property rights associated with water ownership have adopted rules that are rooted in historical patterns of water consumption, some of which are excessively wasteful or less productive compared to other alternatives. As a result, paradoxically, in many areas water supply is both abundant and scarce at the same time! It is abundant, because hydrological assessments indicate there are huge amounts of renewable water resources available (see Table 2: 1). It is scarce, because, under the current laws, in many parts of the world water authorities cannot access the water in sufficient quantity to satisfy the demand. What is needed, therefore, are laws that delineate water rights and specify the terms under which water rights may be transferred (Rodney Smith 1988: 12) in such a way as to enable managers to respond to market signals more flexibly. By providing this legal framework, the relationship between effort and reward will become more certain, and everyone will experience the consequences of their actions. At present, the lack of appropriate legal arrangements creates incentives for waste and misuse.

In cases where the issue of 'water rights' has been settled, new options have allowed for reallocating existing water, including water marketing and sales or leasing of water rights, as many commentators have pointed out. Among them, Gleick (1994a: 37) refers to recent experience in California with water marketing and 'banking', and he notes its applicability in the Middle East. He claims that under 'appropriate institutions and incentives' during a long-term drought in 1991 'the California Water Bank bought nearly 1000 million cubic meters of water from farmers at a price of $0.10 per cubic meter. This water was then sold to urban centers, and a small portion was set aside to aid threatened ecosystems'.

The establishment of appropriate institutions and incentives is, undoubtedly, dependent on the settlement of water rights. However, we must not underestimate the difficulties of settling water rights. As Allan (1996a: 82) points out, existing water management institutions attempt to preserve 'the interests of those who have achieved access to water'. Although this 'access to water', which is transformed into 'perceived rights', is increasingly contested by new users, those who already have access to water are determined to retain their dominance. They rationalise this dominance because they perceive local and national economies to be dependent on the current patterns of water usage, regardless of the real contribution of such activities to national economies. At the local and national levels, governments are, in theory at least, available to intervene in protecting a common resource, and indeed, where water is scarce at local level, government control of water allocation is typical. However, water bodies which are within more than one jurisdiction represent an important area of concern. Experience with some international watercourses shows that most of the so-called international 'water conflicts' are, actually, disputes over water rights or perceived rights. According to Clarke (1991), the absence of international agreements concerning water rights of each riparian state, which is a common problem in developing countries, increases the potential for war. He argues that:

> All countries that rely on water originating outside their territory are dependent, in the absence of treaties, on the goodwill of upstream countries. Goodwill can not be guaranteed, especially

under conditions of scarcity. Downstream countries also often find themselves in a weak position to negotiate formal arrangements. This insecure situation is intensifying as water scarcity increases, and control over the national water supply often becomes a political issue, particularly in countries already impoverished by the over-exploitation of natural resources. In contrast, countries that control the water resources of neighbouring nations can wield formidable power. As water has become increasingly scarce, countries have not hesitated to take action to ensure their own supply, even at the expense of downstream countries. Such actions make international disputes inevitable and [increase the] likelihood of armed conflict. (Clarke 1991: 93–4)

Other observers put forward the same argument. Noting that where water is a transnational resource, there are severe problems of political jurisdiction, they argue that the absence of clearly defined property rights concerning shared water resources often damages political relations between countries. Deterioration of political relations, in turn, makes it more difficult to establish regional water agreements and cooperative management systems. Accordingly, they conclude, the lack of settled water rights is the root cause of the water crisis, and a frequent source of dispute. 'Indeed,' Lipschutz (1992: 12) claims, 'the perception that water rights are inequitably distributed, or may prove to be a problem in the future, could be a greater incentive to conflict than the actual supply situation.'

In answering the question 'Is there any apparent or real connection between water availability and violent conflict?', Lipschutz (1992: 12–13) provides some data about the per capita availability of water in 'water rich' and 'water poor' countries, and the extent to which water availability in a select group of countries is dominated by an international river basin. He concludes that 'the correlation between either per capita water availability, or the degree to which water resources are shared internationally, and the potential for conflict, is weak, at best'. He points out, for example, that 'Indochina is water rich but racked by conflict. The Netherlands is water poor, but very peaceful. Both are located in international river basins'. Lipschutz's explanation is an economic one – that 'instability related to water is more closely linked to the general level of economic development within countries'. But another factor needs to be taken into account – the legal factor. An important reason for water conflict in Indochina is the lack of legally-based water agreements. By contrast, in the

Five Approaches to Water Scarcity 35

Netherlands there is a clear legal basis for water rights (see Blom 1992). This point is important because, as we move from local to international resources, international law is necessary to achieve the peaceful resolution of problems and advance prevention of conflicts. But, according to some commentators, there is a paucity of international law on the subject of water rights (Krishna 1995: 33).

Given that, without clearly defined property rights, water resource development is severely constrained, one may ask: why has the evolution of international law in the field of transboundary water resources not yet created conditions conducive to cooperative strategies? The answer is that in many areas, international law *has* created conditions conducive to cooperation, albeit on a limited regional basis. International treaties concerning shared freshwater resources extend back centuries. Today, well over 2000 joint agreements regulate the conduct of users in almost 240 river basins which are shared between two or more countries (Blake et al. 1995: xiv). Out of these agreements, more than 286 international treaties, dealing specifically with freshwater issues, have been signed by nations (Vlachos 1990: 186; Frey 1993b: 58; Naff 1994b: 272), attesting to the possibility of peaceful water conflict resolution.

Some observers advocate the establishment of a 'Global Water Authority' (Bob Bailey 1996). Although this idea sounds promising, developing such an authority would be very problematic because there are many unsettled legal questions, technical complexities, and complicating regional factors which affect international politics. For example, factors which affect the successful negotiation and implementation of an international water agreement by a nation include whether the nation is upstream, downstream, or sharing a river as a border; the relative military and economic strength of the nation; and the availability of other sources of water supply (Gleick 1994a: 39). As a result of these complexities, different legal doctrines have been developed concerning the criteria for defining water rights of riparian nations of a shared water basin. In the following section we shall briefly review these different doctrines.

Doctrines of water rights

Five distinct theories relating to the governing principles of water law can be identified:
1 absolute territorial sovereignty
2 absolute territorial integrity
3 community of property in the waters

4 limited territorial sovereignty
5 optimal development of the river basin.

The doctrine of *absolute territorial sovereignty* is based on the theory that it is a sovereign right of each state to make full utilisation of all water resources flowing within its territory, irrespective of the effects beyond its boundaries. The concept of absolute sovereignty is associated with the Harmon Doctrine, named after a statement by US Attorney-General Harmon, relating to a dispute in 1895 between the USA and Mexico over the use of the Rio Grande (Thomas and Howlett 1993: 16).

The doctrine of *absolute territorial integrity* emphasises the integrity of the river basin. No riparian can change the natural flow of the river. The concept of absolute territorial integrity is advanced, among others, by Oppenheim, who stated that:

> It is a rule of International Law that no State is allowed to alter the natural conditions of its own territory to the disadvantage of the natural conditions of the territory of a neighbouring State. For this reason a State is not only forbidden to stop or to divert the flow of a river which runs from its own to a neighbouring State, but likewise to make such use of the water of the river as either causes danger to the neighbouring State or prevents it from making proper use of the flow of the river. (1948: 430)

The doctrine of *community of property in the waters* asserts a reasonable share or equitable use by all riparians, not causing unreasonable harm to any other riparian. Henry Farnham was the principal exponent of this theory, arguing that international watercourses were the common property of all the states through which they flowed, and that no one state should intervene to diminish the resource for others sharing it (Thomas and Howlett 1993: 16).

The doctrine of *limited territorial sovereignty* restricts state sovereignty and ties the riparian states to share water resources according to such criteria as historic use, arable land, and population (Frey 1993b: 58). According to this view each riparian state, regardless of whether an international watercourse originates in or traverses its territory, has a vote in deciding what measures are adopted within the watercourse as a whole. However, the state which has been using the water the longest has some priority (Wolf and Dinar 1994: 81; Thomas and Howlett 1993: 17).

Finally, the doctrine of *optimal development of the river basin* envisages the development of the basin without regard to national boundaries. This theory involves contestable notions such as 'optimal', 'reasonable', and 'equitable', over the choice of allocational criteria. It also presupposes basin-wide institutions that may not exist (see: Falkenmark 1986a: 108; Frey 1993b: 58; Wolf and Dinar 1994: 70).

Although riparians who make contentious claims over shared rivers rarely resort to litigation in international courts of law, they always assert the particular legal theory which best justifies their demands, using it more as a bargaining ploy than as an objective, detached legal argument. Hence, for example, downstream states reject the doctrine of absolute territorial sovereignty, while upstream states reject the doctrine of absolute territorial integrity. The basic problem in dealing with such differences is the absence of binding treaties to govern the general and specific terms of shared waters, and the lack of essential inter-riparian or international institutions to assure compliance among the users (Naff 1994b: 272).

However, more than one hundred treaties restrict the total freedom of action of upstream nations (Gleick 1994b: 47). Such treaties are a prerequisite for transforming legal theory into the institutional application of law. They may be basin-focused agreements which deal only with a particular river basin or universal treaties such as the United Nations Convention on the Law of the Sea (UNCLOS) (UN 1983) and the Convention on the Law of the Non-Navigational Uses of International Watercourses (UNCLNUIW) (UN 1997). Basin-focused treaties are traditionally a practical arrangement by which the riparian countries can bring together a set of effective legal instruments for preventing and solving disputes that arise over shared water resources. These agreements often provide for the establishment of joint river commissions, which in some cases have merely an advisory function, but in other cases have decision-making authority. Accordingly, the achievements of joint river commissions vary greatly in different river basins (see: Fox and LeMarquand 1979). River commissions for the Rhine (Schulte-Wulwer-Leidig 1992), the Senegal (Haddad and Mizyed 1996: 10), and the Indus (Alam 1998; Mehta 1986) provide examples of well-functioning commissions with decision-making control.

Supranational institutions with the authority to create general obligations which are binding on member states have proved to be effective in managing international river basins for the interests of the riparian communities, and legal experts have been actively engaged in the formulation of practical legal principles governing shared freshwater resources.[3] So far, two international legal codes have been drafted, though neither is binding yet. The first was produced by the International Law Association at its 52nd Conference in 1966 – the Helsinki Rules on the Uses of the Waters on International Rivers (ILA 1967). The drafting of the Helsinki Rules represented a systematic attempt to codify international water law. The Rules embody the notions of the 'reasonable' and 'equitable' sharing of water resources and recognise the 'international drainage basin' as a fundamental concept for formulating international water law (Article IV). However, the contestability of the concepts of 'reasonable' and 'equitable', and the controversy over the definition of an 'international drainage basin', has prevented unanimous acceptance of the Helsinki Rules as the legal basis for dealing with international watercourses (Thomas and Howlett 1993: 17).

The second legal code has been progressively formulated by the International Law Commission (ILC), which is a United Nations affiliated body, since 1970 when the UN General Assembly charged the Commission with the task of 'progressive development and codification of the law on water courses for purposes other than navigation' (UNGA 1970). Recognising the significance of international watercourses, after 11 years of consideration, the United Nations delegated the codification of international water law to the ILC. After two decades of deliberation, the Commission concluded the drafting and provisional adoption of 32 articles on the law of 'the Non-navigational Uses of International Watercourses' (ILC 1991). Among the suggested general principles were those of equitable utilisation, the obligation not to cause harm to other riparian nations, and the obligation to exchange hydrologic and other relevant data and information on a regular basis (Gleick 1994a: 39). Despite considerable delay in completion of its task, which evoked pessimism in observers such as Naff and Matson (1984: 158), Falkenmark (1986a: 108) and Frey (1993b: 58), the Commission presented a final set of 'Rules on the Non-navigational Uses of International Watercourses' to the UN General Assembly in late 1994, and requested that the General Assembly convene a plenipotentiary conference for elaboration of a convention on the basis of the proposed draft. Thereafter, a working group in the Sixth Committee (Legal Committee), was designated by the General Assembly to finalise the draft in accordance with the comments, recommendations, and concerns of member states. The working group, after considerable elaborations, submit-

Five Approaches to Water Scarcity 39

ted the text of the draft convention (a 37-article draft framework convention and a 14-article annex).[4] Eventually, on 21 May 1997, the UN General Assembly adopted a resolution, with a vote of 103 in favour, 3 against, and 27 abstentions, to which the Convention on the Law of the Non-navigational Uses of International Watercourses was annexed.[5] This convention is a comprehensive global code of law to govern the management of internationally shared water resources – a promising legal structure for settling international disputes over water resources which invites all states and regional economic integration organisations to become parties to it. The convention is important in that it aims to shift international water disputes from contests of power to considerations of fair rights and mutual obligations. Inherent in its rules is the responsibility of each state to use water efficiently and to avoid depriving or damaging a co-riparian state.

The adequacy of international law

The search for establishing 'legal' rules to manage water use is thus not new (see also Chapter 1). This is because, without appropriate legal channels and adequate rules, the potential for conflict over water increases both at local and at international levels. To avoid conflict where water resources are shared, upstream and downstream users must agree on how water is to be allocated. It is also well recognised that lack of agreement between riparians makes the financing of projects much more difficult (Krishna 1995: 30–31). Moreover, the scarcer the resource, the more sensitive are the parties towards any scheme that appears to change the status quo. Therefore, many conflicting claims have to be balanced and reconciled before any plan of water resources development can emerge in a transnational basin. Looking at the issue from a legal perspective, only with the formation of international agreements – bilateral, multilateral or universal – can there be created an adequate array of effective legal instruments for preventing and solving disputes which arise over shared water resources. In other words, law is regarded as the major determinant in finding and maintaining legitimate and sustainable solutions.

However, international law *per se* cannot serve as a complete solution to the water crisis because, as Dellapenna (1995: 58) points out, while international lawyers can formulate advanced legal arrangements, their formulas must be translated into effective institutions for the management of transboundary relations, and this depends upon political settlements, a task for diplomats and politicians. If, however, political actors realise the benefits of enforceable accords, sacrifice short-term benefits for long-term and sustainable achievements, and overcome the complexities mentioned above, the successful negotiation and implementation of international law will certainly play a crucial part in reducing the risks of water conflict.

Water as a technical issue

By contrast to the legal approach, the fourth hypothesis puts its trust in technical ingenuity rather than legal codification. In doing so, it also contrasts with the economic approach, in that while most economists would agree that the increasing scarcity of water is an allocation problem, not a supply problem, technological optimists maintain faith in the ability of science, technology, and human ingenuity to accommodate the world's population demand for water (Narveson 1997). They argue that there is more than enough water on the earth to satisfy all the water needs of humankind (Mastrull 1995: 36); the problem is whether appropriate technology is available or not. This technological optimism is challenged by pessimists who predict a future of Malthusian pestilence, famine, and war, all caused by rising population pressure (Papp 1991: 519). But technological optimists claim that history shows that when demand grows, people will find a suitable skill, technique, or method to solve the deficit problem. The optimists accuse the pessimists of underestimating the ability of humankind in producing the complex scientific, economic, and legal climate in which technical and social ingenuity can flourish. Modern civilisation, they claim, is capable enough of inventing the necessary technologies and of initiating the essential management skills to implement these technologies. On this view, almost no place is too parched or remote to reach with an adequate supply of water. If there appear to be severe local water crises, these are due to technological under-investment or inadequate water management know-how (Mastrull 1995: 36). In other words, the problem is not scarcity of water, but rather the scarcity of investment for technological facilities and/or the scarcity of managerial skills permitting more comprehensive and efficient use of the existing water. Let us briefly explain these two factors – technology and management techniques.

Technology

Technological optimists argue that since 1700 BCE, when Babylonians employed windmills to pump water for irrigation, technological feats have literally made deserts bloom in different parts of the world. To name just a few recent examples, the California Aqueduct in USA, the Aswan High Dam in Egypt, the Karakum Canal in Central Asia, the Great Man-Made River in Libya, the GAP project in Turkey, the controversial Three Gorges Project in China, huge desalination plants in the Arabian Peninsula, and advanced water conservation technologies in Israel, have all been hailed as 'hydrological miracles'. Similarly, pollution can be curbed greatly by technological means; even highly contaminated water can be made drinkable again. Just as green revolution technologies greatly increased the production of grain for domestic consumption and of cash crops for export, and scientific developments in biotechnology and genetic engineering promise to curtail even more so the limits of natural selection and overcome food scarcity, so the prospect of eliminating water scarcity by technological innovation seems limitless.

Most alluring of all is the possibility that there are great undiscovered reservoirs throughout the globe. With the arrival of space shuttles, we are now able to trace these reservoirs through satellite photography. For instance, in 1990, at a water summit in Cairo, Farouk El-Baz of Boston University announced that remote sensing data indicate there may be twice as much water stored underground worldwide as previously assumed. He raised hopes among African nations by revealing that there are untapped supplies of subterranean water in the driest part of the Egyptian Sahara (Linden 1990). It is estimated that the ground water stored in aquifers deep beneath the Sahara would supply anticipated needs for some 200 years. If these were used appropriately, many new areas could be brought under cultivation, enabling famine-prone countries such as Egypt and Sudan to become bread baskets for the Middle East (see: *Washington Times*, 26 February 1997; Suliman 1992; Hillel 1994: 10). This notion has guided the cultivation of barren lands of the Negev Desert, a region that was written off as largely uncultivable for millennia (Elmer-Dewitt 1989). Some hydrogeologists are firmly convinced that there are plenty of usable but untapped water resources in deep aquifers under the Saharo-Arabian deserts which could serve as a safeguard against scarcity and strife. They argue that what is most needed today is for the international parties to

42 *Water Politics in the Middle East*

Figure 2.1 Technological options for managing water scarcity

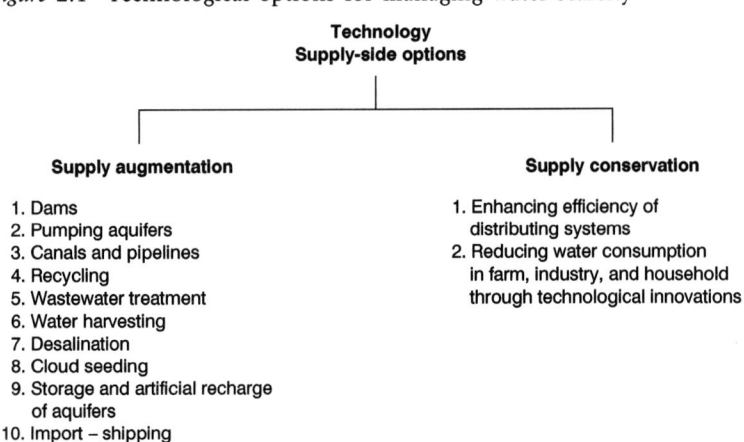

Technology
Supply-side options

Supply augmentation
1. Dams
2. Pumping aquifers
3. Canals and pipelines
4. Recycling
5. Wastewater treatment
6. Water harvesting
7. Desalination
8. Cloud seeding
9. Storage and artificial recharge of aquifers
10. Import – shipping
    – towing icebergs
    – transnational pipelines

Supply conservation
1. Enhancing efficiency of distributing systems
2. Reducing water consumption in farm, industry, and household through technological innovations

sit together, confer and come up with new ideas and new methodologies in water management. If these untapped water resources can be developed with new methodologies and used efficiently, then we have another resource that is much more economical than seawater desalination.

The variety of technological options for managing water scarcity is illustrated in Figure 2.1. With tools and technologies available today, vast water supplies are increasingly provided through dam-building,[6] pumping renewable aquifers, mining fossil water of deep aquifers, expanding efficient water distribution systems, desalination of sea water and brackish water, cloud seeding, recycling and wastewater treatment. In addition to the idea of importing water via transnational pipelines, other non-conventional water resources alternatives including transport by large tankers, towing of large capacity rubber bags, and iceberg towing are contemplated. Moreover, as Postel (1992: 99–160) shows, with the technologies available today, enormous water savings are possible in agriculture, industries, and cities. Agriculture could cut its water demands by up to 50 per cent, industries by up to 90 per cent, and cities by 30–5 per cent with no sacrifice of economic output or quality of life. Postel believes that advanced technology, conservation, efficiency, recycling, and reuse of water could 'get us through many of the shortages on the horizon, buying us time to develop a new relationship with water systems and to bring consumption and population growth down to sustainable levels' (Ibid.: 191). If that were to happen, people in water-scarce areas could increase their effective water resources to meet their needs without any heightening of conflict.

Management techniques

Although technology has proved to be a decisive factor in water supply management, it would be seriously misleading if we perceive water scarcity merely in hydrological and technological terms. Given that sufficient capital is available, the successful implementation of technology, especially in the developing countries, presupposes three conditions: (1) that water scarcity is the real bottleneck to growth; (2) that institutions exist which can *manage* the technology; and (3) that the technology fits the local social structure and values. According to the World Resources Institute, across the world, 65 to 70 per cent of the water people use is lost through management inefficiencies (Linden 1990: 58). There are innumerable examples of water scarcities resulting from inappropriate resource management policies. For instance, Clarke (1991: 50) claims that the Cheerapunji region of India has a world record of rainfall – 9000 millimetres per year – yet we can call it a 'desert', because heedless deforestation has removed the region's ability to preserve water. Hence, despite an annual rainfall of more than nine metres, one of the wettest regions of the world now suffers from water scarcity. This and many other cases of deforestation, soil erosion, salinisation, waterlogging, and pollution of water resources highlight the importance of resource management. The challenge now is that, according to Viessman (1990), Postel (1992) and Homer-Dixon (1994b), society must put as much human ingenuity into managing to live in balance with water as it has put into controlling and manipulating it.

As discussed earlier, there are many economic and legislative leverages by which water authorities can manage – that is, regulate and control – the patterns of water supply and its consumption in each society. A detailed discussion of the various merits and drawbacks of each of the options which are illustrated in Figure 2.2 is well beyond the scope of this study, but there is a large body of literature which brings out the lessons and pitfalls of past water management experience in different countries,[7] and underlines the necessity for countries to learn from one another's experience and to pool efforts to address the problem of water scarcity.[8] In particular,

Figure 2.2 Managerial options for managing water scarcity

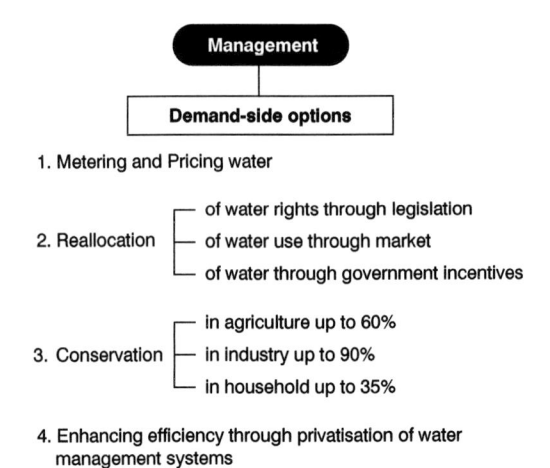

```
                    ┌──────────────┐
                    │  Management  │
                    └──────┬───────┘
                    ┌──────┴────────────┐
                    │ Demand-side options│
                    └───────────────────┘
```

1. Metering and Pricing water

2. Reallocation
   — of water rights through legislation
   — of water use through market
   — of water through government incentives

3. Conservation
   — in agriculture up to 60%
   — in industry up to 90%
   — in household up to 35%

4. Enhancing efficiency through privatisation of water management systems

many writers[9] argue the case for the greater use of 'demand management', whereby better use is made of existing water supplies instead of investing in new supply capacity to satisfy increased future requirements.

Since water is a renewable resource and, in principle, can be reused several times, 'the availability of water for human use depends as much on how it is used and how water resources are managed as on any absolute [hydrological] limits' (World Resources Institute 1996/7, ch. 13). Indeed, as Homer-Dixon (1992, 1994a, 1994b) notes, technological progress to increase the absolute supply of water may be neutralised by management failure. For example, large-scale projects often change the relative value of water resources, alter instituted property rights governing resource distribution which have been fairly effective though not necessarily optimally efficient, and transform social relations. As a result, they could concentrate water resources in the hands of a few people and subject the rest to scarcity. The same is true in the international arena; engineering technologies now give certain nations the tools to overpump aquifers and redirect rivers for their own national purposes, thereby changing the relative value of water and its traditional pattern of distribution and in many cases producing local water scarcity.

On the other hand, as the latest report of the World Resources Institute (1996/7) points out, implementing appropriate water management policies can stretch even scarce water supplies much further (see Figure 2.2). The problem is that, while technological development proceeds rapidly, managerial reform only creeps, and despite breathtaking advances in technology, the development of water management policies is often impeded by an array of institutional constraints. According to Viessman (1990), bringing about institutional change is 'tedious and painful' because it affects special-interest groups, agencies, statutes, laws, regulations, social customs and traditions, many of which are 'relics' and hostile to modification.

To overcome these obstacles, therefore, we need to explore alternative modes of resource management, identify the merits and flaws of different options, clarify the advantages that could result from change, and convincingly explain the efficacy of the proposed changes. Fortunately, the international community is at the brink of this transformation. It is not only well recognised that inefficient management of water has created scarcities which affect many environmentally sensitive regions, but it is also commonly accepted that we need to create management systems that buffer people from the effects of scarcity.[10]

The literature indicates that, traditionally, the reaction to water scarcity has been to focus on how to increase the supply. For example, for half a century, the World Bank and other international organisations provided money and expertise to augment the supply by building facilities and transferring technology. Increasingly, however, these organisations now see their task as providing support in the formulation of new water resources management strategies (Le Moigne 1994a: 15). When it seemed to be a problem of supply, the World Bank spent more than $36 billion on building megadams, canals, pipelines, and other supply augmentation projects, but now it argues that the water scarcity in most countries stems mainly from the way water is used (World Bank 1995: 8). Hence, despite the fact that the Bank estimates that Middle Eastern and North African (MENA) countries will need to spend $60 billion on water projects in the next 10 years to increase their water supplies by 50 per cent and avoid disputes which many commentators fear could cause another war in the region (Ibid.: 26), nevertheless, it emphasises that the solution to water scarcity lies less in supply than in demand management. The Bank suggests that countries should turn from supply-side projects and start to invest in demand-side plans. To encourage this process, the Bank will issue loans for projects such as drip irrigation, leak-plugging and sewage treatment.

There are two major advantages in this shift: first, demand-side solutions are cheaper than supply-side options. Second, demand-side management is more efficient than supply-side management in enabling societies to avoid violence when confronting water scarcity. According to Homer-Dixon (1990, 1991a, 1994a, b, 1995), in the case of interstate or domestic conflicts induced by resource scarcity, societies which can adapt by demand-side measures are more able to avoid strife. Such countries either continue to rely on their indigenous resources by using them more sensibly, or 'decouple' themselves from dependence on their own depleted or limited resources by producing goods and services that do not rely heavily on those resources.

The adequacy of technology and management techniques

New technologies and advanced management techniques thus have much to offer for achieving water security, but the question remains: how far will they take us? Can they give people sufficient access to and control over nature's supply to avert shortages and avoid conflicts for ever? In answering these questions, we have to point out that there are some weaknesses in the technological approach. First, as Homer-Dixon (1991a) notes, we cannot take it for granted that human scientific and technical ingenuity will always overcome all types of scarcity. Second, scientific and technical knowledge takes time to develop, and its practical application to society often takes more time. Therefore, technical solutions to environmental scarcity may arrive too late to prevent catastrophe. Third, we cannot accurately foresee the consequences of our technological adaptation and intervention strategies. The experience suggests that, although technology has substantially helped people to gain access to nature's supply in the short term, it has often had negative long-term consequences. Thus, if humans depend only on technological innovation to overcome an increasingly hostile environment which is of their own making, a real danger exists that the same technological innovation will add to environmental degradation as it seeks to overcome it (Dabelko and Dabelko 1995). For example, as Postel (1992: 18–24) points out, building ever more and larger projects to meet soaring demand but giving little attention to the ecological consequences of losing rivers, lakes, aquifers and wetlands in the process, is 'masking' scarcity rather than resolving it. Moreover, where demand grows faster than efficiency measures can release further supplies, technology alone cannot avert shortages and conflicts.

Fourth, even with unlimited water resources, we may not be able to stop global warming and its climatological and hydrological effects. This and other environmental crises that are being faced today have attained such a critical level that humans are forced to confront certain basic questions about their relationship to the environment. We cannot remove these threats to our milieu simply by innovating more advanced technology and initiating up-to-the-minute management policies. The whole environmental crisis, a part of which is the water crisis, represents not only a technological, but also an ideological and social crisis of modern industrialised civilisation (Van Dijk 1994; Dolatyar 1996; Colin Ward 1997). It is these crises which gave birth to environmentalism as a new philosophical and political discourse,[11] and it is to the environmental hypothesis that we now turn.

Water as an environmental issue

Environmentalists (greens) suggest that the water crisis is part and parcel of the global environmental crisis that we are facing today. They argue that the earth is a finite planet with limited resources and capacities. This finitude, and the scarcity it implies, places limits on our use of its resources beyond which the environment fails to recover and restore its equilibrium. This is true regarding every eco-geographical region as well as every natural resource which it provides – including fresh water. Introducing the notions of 'the limits to growth' and 'sustainable development', and preaching a new 'environmental ethics', greens maintain that the global environmental crisis and its components such as the water crisis stem directly from our erroneous intellectual approach towards the natural world and our misguided behaviour in relation to the environment.

The environmental approach to resource scarcity takes into account the security, economic, legal, and technical dimensions of the issue along with special emphasis on the interrelations and interdependence between different parts of the natural world. From this holistic perspective, the natural world is seen as an interdependent system in which the survival of each unit depends on the viability of every other unit. We cannot separate the units and study them in isolation; rather we must study their interrelationships. Greens argue that a proper understanding of this interdependence and

48 *Water Politics in the Middle East*

reciprocal influence will guide us in formulating our relationships in the natural world with sensitivity to the environment. This view of reciprocity and interdependence has greatly influenced the greens' prescriptions for economic, political and social arrangements in the human world.

Before discussing the implications of this approach for the water scarcity issue, it should be noted that the extensive literature on environmentalism indicates that the desire to protect ecosystems and conserve resources is a manifestation of deep-rooted values which underpin this newest offshoot of political philosophy. It is now well recognised that 'at its heart environmentalism preaches a philosophy of human conduct' (O'Riordan 1981: ix). It should also be mentioned that the rise of the 'environment' as a social and cultural phenomenon has now so penetrated the common consciousness that 'not only marginal social groups but also political parties, industrialists, religious leaders, scientists of all descriptions, even the legal professions, all seek to reflect a sensitivity to "environmental" priorities' (Grove-White 1993: 18).[12] Although a closer look reveals that the differences between those who show a concern for the environment are often greater than the similarities, they generally subscribe to fundamental notions of environmentalism such as 'the limits to growth' and 'sustainable development', which we now examine.

The limits to growth

Greens argue that the Earth is a *finite* object with limited capacities – a limited productive capacity for resources, a limited absorbent capacity for pollution, and a limited carrying capacity for population. The Earth's finitude and its limited capacities suggest that *scarcity* is a fundamental principle 'rooted in the biophysical realities of a finite planet, ruled and limited by entropy and ecology' (Irvine and Ponton 1988: 26). Greens' fundamental commitment to the principle of scarcity as 'an insurmountable fact of life' underlines their argument that there are *natural* limits to economic and population growth (Dobson 1990: 80). Bunyard and Morgan-Grenville, for example, argue that if we continue to use the earth's nonrenewable resources and to abuse its renewable resources at the present rates, it is an inevitable fact that 'at some stage in the future the whole system is going to fall apart' (1987: 327).

From this perspective, consistent and unlimited growth cannot be achieved by resorting to technology, because 'technological solutions cannot provide a way out of the impasse of the impossibility of aspiring to infinite growth in a finite system' (Dobson 1990: 76). This is not to deny that the application of technological solutions facilitates and protracts the course of economic and population growth; the point is that, as Meadows *et al.* (1983: 141) assert, it cannot remove the ultimate limits to that growth. Moreover, a special concern of greens is that all economic activity involves transforming the natural world and this often leads to excessive environmental degradation. This degrading transformation, which is primarily induced by the desire for economic growth and largely facilitated by technology, harms people. But the people who suffer from the damage may be different from those who enjoy the benefits of the growth. The sufferers may be today's poor, or future generations who inherit a degraded environment. This fact provides additional grounds for rethinking our notion of *growth*, and this is the basic reason why greens believe that profound changes in our economic practices, values, social habits and political behaviour need to take place.

What has happened in the water domain during the last three decades substantiates this argument very clearly. By using technology, new water supplies have taken some pressure off rivers, lakes, or aquifers and have provided a temporary release from water scarcity in many parts of the world. However, if societies carry on with business as usual, this bounty will only postpone the day of reckoning for humans and all other species. As we shall see in Chapter 3, the imperative desire for economic growth and the single-minded pursuit of technological development has led almost all countries to disrupt the hydrological cycle on which the renewability and sustained availability of fresh water depends, and to pollute their water systems by agrochemical, industrial and urban wastes. The factual analysis of Chapter 3 shows that the shortage of clean drinking water, floods and drought as a result of deforestation and mining in vulnerable catchments, disruption of food systems by cash crop production, and the ecological impact of large irrigation systems are now common problems in both industrial and developing countries in many parts of the world. These problems indicate that the current engineering-orientated approach to water resources development cannot solve the water crisis and that a new strategy must be adopted to tackle the problem. This new strategy, from the green perspective, is part of a *sustainable development code,* to which we now turn.

Sustainable development

As we saw above, the notion of 'the limits to growth' is the main rationale for greens to argue for *sustainable development* and the need to create a *sustainable society* in which economic growth and development can continue without environmental destruction. Defining the concept of 'sustainable development' has proved difficult and requires us to distinguish between its dark-green and light-green manifestations. To determine what is 'sustainable development', we must at first decide: what should 'sustainable development' sustain? Is it 'human life' which should be sustained or should we go beyond human-instrumentalism and argue that 'the natural environment' has an intrinsic value that should guarantee its 'right to life'? There are complex issues behind each of these statements which are beyond the scope of this study (see Dobson 1990: ch. 2). For the sake of our argument, however, we must at least arrive at a general understanding of the term.

In the economic sense, 'development' is about 'improving the well-being of people' (World Bank 1992a: 34) but sometimes this is regarded as equivalent to 'economic growth'. Raising living standards and improving the quality of life are essential components of development and economic growth is an essential means for facilitating development, but in itself economic expansion is a highly imperfect synonym for development because of the loss of human health and well-being which follows the environmental decline often consequential upon economic and/or population growth. The term 'sustainable development' was brought into common usage by the World Commission on Environment and Development in its report (*Our Common Future 1987*) to unify 'environment' and 'development' and establish a new approach to economic growth. The Commission's definition of the term – meeting the needs of the present generation without compromising the needs of future generations – which is widely accepted, implies that meeting the needs of today's poor is an essential aspect of sustainably meeting the needs of subsequent generations (World Bank 1992a: 8).

Thus, we can say that sustainable development entails a wholly new approach to economic growth. It drives us toward sharing limited resources equitably, using resources efficiently, and developing environmentally sound technologies. It also requires us to harmonise our economic goals and industrial growth with ecological criteria and to modify our goals and priorities accordingly. The deeper basis

Five Approaches to Water Scarcity 51

of this approach is that 'steady and sustainable development is impossible unless the bits and pieces which make up the environment – soil, water, forests, wildlife, etc. – are used in a sound and sustainable manner' (Timberlake 1984: 128).

It should be noted here that the policies, laws, and practices that shape water utilisation today rarely comply with the three basic tenets of sustainable resource use – equity, efficiency, and ecological integrity (see Chapter 3). The major changes that are required in the way water is valued, allocated, and managed today would require a fundamental restructuring of economic activities and reordering of priorities. For instance, as Postel (1992: 187) points out, in a water-scarce region, sustainable development entails raising water's productivity – getting more value out of each unit used while leaving enough in rivers, lakes, and aquifers to keep natural systems functioning well. This means that overemphasising the goal of 'food self-sufficiency' would have to be abandoned in some countries (see Chapters 4 and 6).

Turning the concept of sustainable development into policy also raises a fundamental question about the value and status of the environment. On this question, light-greens suggest that we have to care for the environment because it is in our interest to do so – it helps to sustain human life. By contrast, deep-greens suggest that the environment has an intrinsic value which entitles it to existence regardless of the interest of human beings (Bunyard and Morgan-Grenville 1987: 284). This deep-green notion provides the basis for environmental ethics – a new ethics which is essentially different from an ethics developed from within the traditional paradigms such as 'liberalism', 'realism', and 'Marxism'. First, the liberal, realist and marxist paradigms treat the environment as an external factor which merely provides resources for the economic process, rather than being important in its own right (Williams 1996: 48). Second, underlying the realist, liberal and marxist perspectives is a technocentric and anthropocentric approach to the environment. Liberalism, realism and marxism share a belief in material growth and technological optimism (Eckersley 1992: 21–6), while the environmental ethic opposes anthropocentric values and does not believe in a 'technological fix'. A central feature of this ethic is that it questions the anthropocentric concept of humanity's 'domination of nature' which has long persisted as a principle in Western philosophy, regarding it as the root cause of environmental degradation. Deep-greens insist that 'this ultimately destructive belief must be rooted out and replaced with a biocentric philosophy' (Porritt, quoted in Dobson 1990: 31).[13]

Concerning the economic, political, and social changes which are possible and desirable within the framework of sustainable development, light-green and deep-green responses are different in terms of both the *kind* and the *amount* of change that they prescribe. According to Dobson (1990: 13, 35, 205) and Grove-White (1993: 20), light-greens believe that by resort to the market economy, persuasion and regulation, international agreements, and technological innovations, environmental problems can be solved without fundamental changes in present values or patterns of production and consumption. In terms of change in human relationships with the non-human natural world, light-greens are interested only as far as it might affect the well-being of humankind. Deep-greens, however, believe that the environmental crisis cannot be adequately addressed either within narrow disciplinary boundaries or within the troubled paradigms of state-centric nationalism and human instrumentalism. They are generally suspicious of technological solutions to environmental problems and argue that care for the environment requires radical changes in our intellectual relationship with the natural world and the practices that stem from it (Dobson 1990: 37).

Drawing on aspects of both deep-green and light-green perspectives, the environmentalist approach has shifted the debate over water scarcity away from concern about *physical limits* to growth of supply, toward concern about changing *human behaviour* (see Attfield 1983; Christopher Ward 1997). Environmental ethics, which is a principal part of green politics, explains that, besides natural limits, there are also social and ethical limits to growth (Dobson 1990: 86–8). Although greens are in disagreement over the extent of their ethical community (Dobson 1990: 47–62),[14] the idea of a green ethic has become a familiar part of environmental theory. And this idea has now been incorporated in the literature on water scarcity. For example, writers such as Linda Stark and Sandra Postel talk about the necessity of adopting a new *'water ethic'*, which requires us to protect water's ecological functions, learn to live within the hydrological limits of our ecosystems, conserve water to get as much as possible out of each unit we take from its natural course, share its benefits with others who are in need, and create a society that is sustainable in all respects (Postel 1992: 183–91).

This 'water ethic', which is part of a sustainable development code endorsed by environmentalism, is very different from all the previous approaches to water scarcity rehearsed above. First, it refutes the security perspective by questioning the validity of a narrowly defined concept of 'national security' which is the backbone of the realist paradigm, and introduces instead the concept of 'environmental security' to which we shall return shortly.

Second, it entails a wholly new approach to economic growth, one that harmonises economic goals with ecological capacities. While it welcomes an adjusted market mechanism as a means of giving water a higher value in its economic functions and uprooting wasteful and unproductive uses, at the same time it highlights the failure of the market system to reflect the full social and ecological costs of water, arguing that the economic philosophy which underlies the setting of prices is concerned more with cost-recovery than with signalling resource scarcity. This is why the market mechanism has failed to safeguard the life-sustaining functions of many delicate ecological systems. To overcome this deficiency of the economic approach to water management, the water ethic introduces a set of guidelines and moral responsibilities which play an important role in the quest for sustainability.

Third, while acknowledging the importance of establishing water rights as a prerequisite for tackling the water crisis, the water ethic proposes that the principles of equity and fairness must also be institutionalised in international relations if tensions over water are to be defused. If millions of people die or suffer from waterborne diseases, it is not because of lack of water rights, but rather because of a lack of social and political commitment to meet the basic needs of the poor. Indeed, as Colin Ward (1997) rightly argues, there is a crisis of social responsibility reflected in the water crisis. From privatisation in Britain to the displacement of millions through dam-building in the developing world, water has been wrongly appropriated as a commodity by the powerful, he maintains.

Fourth, while the water ethic encourages the application of appropriate technology and suitable techniques to reduce the amount of water used in agriculture, industry and household, it rejects the idea of a 'technological fix' by introducing the notion of 'limits to growth'. The only challenge now, from this perspective, is: 'to put as much human ingenuity into learning to live in balance with water as we have put into controlling and manipulating it' and 'to bring consumption and population growth down to sustainable levels' (Postel 1992: 191). To achieve this end, the involvement of international organisations and NGOs, the expansion of international economic interdependencies, the evolution of international law in the field of transboundary water resources, and the use of new technologies are all helpful in creating appropriate conditions for cooperation. However, the ultimate solution rests in the environmental consciousness of people and decision-makers. This is because environmentalism emphasises the shared interests of nations in a global civil society rather than the exclusive interests of autonomous rival states in a world of anarchy. By introducing the notion of *environmental security* as an alternative to the traditional concept of *national security*, environmentalism prepares the ground for establishing a cooperative international system.

Environmental security an alternative to national security

By drawing attention to the complex global interdependencies of the late twentieth century, environmentalists argue that grave consequences of environmental degradation and resource scarcity are not confined to *national* borders and will inevitably affect all parties in one way or another. These arguments have provided an opportunity for re-examining the concept of 'security'.[15]

In the traditional discourse of security, the notion of 'security' implies a threat of violent action coming from an imputable agent to which a response can be made. Such a threat, either to the security of a state or to international security, and the reciprocal response generally involves armed force (Johnson 1991: 172). Regarding the new environmental pressures, such as degradation and depletion of international water resources, that transcend national borders and penetrate the boundaries of national sovereignty, some observers such as Jessica Mathews (1989) have suggested that the assumptions and institutions that have governed international relations in the postwar era are a poor fit with new environmental realities, and that we need to 'redefine security'. This view has grown increasingly popular over the past few years and many writers have called for reconceptualising the notion of security so that non-military threats, particularly environmental degradation and resource scarcities, could be incorporated into the modern conception of national security. However, the major defect of this approach is that environmental security itself could become militarised through the co-option of the environmental agenda by a traditional security agenda.[16] In other words, states might be encouraged to respond to non-military environmental threats by military retaliation.

There is another interpretation of environmental security which goes far beyond the limited boundaries of national security. This interpretation is based on a new worldview in which environmental security – as a form of collective security – could be regarded as an alternative to national security, which is a variant of individual security.[17] From this perspective, continuing dependence on the troubled concepts of territoriality, sovereignty, national interest, and state foreign policy, which have historically provided the framework and rationale for military threats and actions, cannot provide a proper response to emerging global changes – especially global environmental change (Hugh Dyer 1996: 23). By contrast, the concept of environmental security would take account of both the universal and the intergenerational scope of the threat, and 'the maintenance of the local and the planetary biosphere as the essential support system on which all other human enterprises depend' (Buzan 1991: 19–20).

Thus, it could be argued that the developing logic of 'environmental security' envisages an international relations system that promotes global relations among regional and local actors rather than focuses on traditional inter-state relations. Adapting a normative approach, Dyer propounds that:

> Environmental security as a value [is] developing into a sociopolitical norm in a global context, one which influences both behaviour (as in political regimes) and knowledge (as in theoretical paradigms). Environmental security arises in a changing international context where interdependence is already widely accepted as the baseline of international relations, and where shared values such as environmental security are more salient than the particularistic interests (such as national politico-military security) of the individual nation states. This transition coincides with the relative decline in the salience of nuclear deterrence and the increasing salience of environmental concerns. In this sense, the environment becomes the manifestation of new political values and norms as the detritus of the Cold War experience and the international system it bolstered is tossed out. (Hugh Dyer 1996: 31)

The merit of taking such a broad approach to security – the security of the entire human environment – is that on this view there are no military solutions to 'environmental security'. Indeed, the resort to war not only *cannot solve* the problem but rather *exacerbates* it. The solution is to understand the 'limits to growth' of each eco-geographical region in which we live and to create a 'sustainable society' accordingly; otherwise, the 'environmental security' of all parties will be in jeopardy. Undoubtedly, one of those resources which are fundamental to life and require particular consideration is fresh water. Properly interpreted, the concept of environmental security sees water scarcity not as a national security issue but as a global security issue – one in which we are all involved and responsible, not least because our development policies – such as on CO_2 emissions – could contribute towards water scarcity in many parts of the world. This is why so much emphasis has been placed on the need to link the environmental and developmental aspects of the water issue in chapter 18 of Agenda 21 (UNCED 1992). As Thomas and Howlett (1993: 19) suggest, there is now an international consensus on the urgent need for integrated management of water resources as a prerequisite for socio-economic development and conflict avoidance in the future.

Conclusion

As indicated at the beginning of this chapter, the discussion of the five major approaches to the water issue leads us to the conclusion that, since water plays a multifunctional role in every aspect of human life, water-related issues need to be seen in the context of the relationship between humans and their living environment as a whole rather than in isolation – as a military, economic, legal, or technological problem. This means that from the perspective of 'environmental politics' the water scarcity issue should be dealt with in the context of *environmental security*, which is very different from the context of *national security* proposed by the traditionally dominant perspective of 'power politics'.

As Boersema (1994: 20) rightly claimed, 'the days when the environmental issues were seen as a technical defect in our social machinery are now in the past. The realisation is dawning that we are confronted with a very complex problem, which is closely interwoven with our civilisation'. And as Van Dijk (1994: 56) suggests, 'besides all the necessary technical, economical and political measures... it is of great importance to ask ourselves something about human nature.... To understand the present crisis, we should become aware of the route which has led us into this crisis'. Regarding the water crisis, the next chapter will demonstrate that, as long as the issue is dealt with from conventional fragmentary approaches, it is, and will remain, an intractable problem. The solution rests in a holistic – environmental – approach.

Notes

1 For example, the following sites on the Internet are available:
http://nn.apc.org/sei.un.water/ (UN/SEI Comprehensive Freshwater Assessment Project)
http://www.uwin.siu.edu/ (University Water Information Network)
http://www.hydroweb.com/ (International Association for Environmental Hydrology)
http://www.waterweb.com/ (Water Web)
http://tornade.ERE.UMontreal.CA/~simardn/Index.html/watertu.htm (Boundary Water Bibliography)
http://www.fao.org/default.htm (FAO Homepage)
http://www.soas.ac.uk/Geography/WaterIssues/ (Water Issues Group at SOAS, University of London)
http://www.ssc.upenn.edu/~mewin/ (Middle East Water Information Network)
http://www.ssc.upenn.edu/~mewin/links.html (Middle East & Water Links)
http://www.mother.com/uswaternews/ (US Water News Homepage)
http://www.fsk.ethz.ch/encop/ Environment and Conflicts Project (ENCOP)
http://www.library.utoronto.ca/www/pcs/database/libintro.htm (The Environmental Security Database).
2 See, for instance, Ben-Shahar et al. 1989: 49; Bulloch and Darwish 1993: 32; Butler 1995: 34; Cooley 1984: 3; Falkenmark 1986a: 89; Gleick 1993a: 85, 1993c: 108–10; George Joffe 1993: 73; Lowi 1993b: 115;. Richards 1992: 9; Thompson 1978: 66–7.
3 See, for instance, Berber 1959; du Bois 1995; Bruhacs 1992; Burchi 1992; Cano 1989; Caponera 1973, 1983, 1992, 1995; Dellapenna 1995; Flint 1995; Garretson et al. 1967; Hayton 1983; Hayton and Utton 1989; ILA 1967; IRAL 1984, 1985; Kiss 1992; Kliot 1995; Krishna 1995; McCaffrey 1993; Mehta 1986; Radosovich 1979; Solanes 1992; Teclaff 1967, 1976, 1977, 1985; Teclaff and Utton 1981; UN 1984; UNWC 1978;
4 The report of General Assembly's Sixth Committee (document A/51/869).
5 General Assembly Plenary–Press Release GA/9248 99th Meeting (AM) 21 May 1997.
6 Thirty years ago, on average 350 dams were built annually worldwide. Although the best sites have now been used, still almost 170 projects are initiated each year across the world to build dams (Lean 1993: 21).
7 See, for example, Abate 1994; Ahiram and Siniora 1994; Al-Alawi and Abdulrazzak 1994; Al-Ruwaih 1995; Al-Weshah 1992; K. Anderson and Holeman 1995; Bakour and Kolars 1994; Ben Meir 1994; Blom 1992; Blomquist 1992; Bodgener 1984; Calvert 1993; Cano 1986; Chomchai 1986, 1995; Choudhury and Khan 1983; Dana 1990; Eaton and Eaton

58 Water Politics in the Middle East

1994; Elmusa 1994; El-Nashar and Qamhiyeh 1991; Gelsse and Arenas 1995; Grove 1985; Guest 1993; Hamdy et al. 1995; Hyde-Price 1993; Jayal 1985; Jellali and Jebali 1994; LeMarquand 1986, 1990; Major et al. 1996; Mumme 1995; Murakami et al. 1995; Schiffler 1995; Temperley 1992; Zaman 1983.

8 See among others: Abdel Mageed 1994a, b; W. M. Adams 1992; Allan 1994b; Allan and Karshenas 1996; Bassett 1986; Biswas 1978, 1983, 1994; J. A. Dyer 1989; El Morr 1995; Falkenmark 1986b; Fleckseder 1992; Fox and LeMarquand 1979; Haddadin 1996; Hartvelt and Okun 1991; Hvidt 1997; Peter Kemp 1995; London and Miley 1990; McDonald and Kay 1988; Murphy and Sabadell 1986; Overman 1976; Shapland 1995; Zarour and Isaac 1993, 1994.

9 See among others: Arlosoroff 1996; Berkoff 1994; Bhatia et al. 1993; Dave 1991; Peter Kemp 1995; Naff 1994a; Postel 1992; Plusquellec et al. 1994; Pisani 1995; Serageldin 1995.

10 See for example, Berkoff 1994; Bhatia et al. 1993; Blake et al. 1995; Dellapenna 1995; Dinar and Loehman 1995; Goldberg 1995; Hartvelt and Okun 1991; Kirmani and Rangeley 1994; Le Moigne 1994a, b; Le Moigne et al. 1994; Munasinghe 1990; Plusquellec et al. 1994; Serageldin and Steer 1994; Serageldin 1994, 1995; Tuijl 1993; Viessman 1990; World Bank 1992b, 1993, 1995; Young 1996.

11 For background and useful discussions of contemporary environmental politics see, among others: Attfield 1983; Bookchin 1986; Brennan 1988; Bunyard and Morgan-Grenville 1987; Caldwell 1990; Capra 1985; Dobson 1990; Hugh Dyer 1996; Eckersley 1992; Elliot and Gare 1983; Warwick Fox 1984; Grove-White 1993; Lee 1989; McCormick 1989; Milton 1993; O'Riordan 1981; Paehlke 1989; Pepper 1984; Pirages 1977; Porritt and Winner 1988; Redclift and Benton 1994;. Spretnak and Capra 1985; Thomas 1992; Tokar 1987.

12 The past two decades have also witnessed a flurry of environmental activity at the international level. Over 70 multilateral conventions or regimes have been negotiated, including the 'Convention on the Protection and Use of Transboundary Watercourses and International Lakes' which entered into force in October 1996, and the most recent – the 'Convention on the law of the Non-Navigational Uses of International Watercourses' which was adopted by the UN General Assembly on 21 May 1997. Although many states have failed to sign these conventions and monitoring and enforcement remain imperfect, a raft of international environmental law now exists to guide and regulate state actions. Moreover, regional organisations as diverse as the Organisation of African Unity, the European Union, the North Atlantic Treaty Organization, the Organisation for Economic Co-operation and Development and Asia Pacific Economic Co-operation have all engaged in some level of environmental activity. Furthermore, the United Nations has acted to incorporate environmental issues into many of its specialised agencies, including the United Nations Development Programme, the Food and Agriculture Organization, the International Labour Organization, the World Health Organization, the International Maritime Organization and the United Nations Education, Scientific and Cultural Organization (Matthew 1995).

13 For detailed discussion of this point see among others, Achterberg 1994; Jacobs 1994; Musschenga 1994; Van Dijk 1994; Vermeersch 1994; Willetts 1996; Zweers 1994.

14 Some writers extend the limits of ethical community to include 'all sentient beings' (Midgley 1983: 89). Others extend it further on the ground that 'all life has intrinsic value' (Bunyard and Morgan-Grenville 1987: 281). Others still argue for an 'intuitive' notion of 'life' which may cover a stream, lake or object (Naess 1984: 202–3). The idea of 'land ethic' expands the boundaries of the community to include soils, waters, plants, and animals, or collectively: the land (Leopold 1949, pp. 203–4).

15 Indeed, environmental issues have become the core of international debate as a key variable in understanding security and conflict in the late twentieth century. Arthur Westing, Lester Brown, Richard Ullman, and Norman Myers are pioneers of this still ongoing debate. The interest of many politicians and academics in the topic has created a lively discussion between those who regard the redefinition of *security* as the starting point to change the faulty international system; those who seek to incorporate the concept of *environmental security* into existing institutions and practices; and those who regard it as a muddled reaction to the general confusion that has followed the sudden end of the Cold War. The debate has contributed to the development of environmental politics as an expanding new political discourse that soon began to affect political agendas at the local, national and international levels. For literature which specifically addresses environmental security, see: Bennett 1991; Lester R. Brown 1977, 1986; Neville Brown 1989, 1990; Dabelko and Dabelko 1995; Dalby 1992a, b, 1995; Deudney 1990, 1991; Deudney and Matthew 1995; Hugh Dyer 1996; Gleick 1989a, b, 1991; Homer-Dixon 1991a, b, 1994a; Imber 1991; Kirk 1991; Mathews 1989; Matthew 1995; Myers 1986, 1989, 1993; Pirages 1991; Renner 1989a, b, c; Rowlands 1991; Schrijver 1989; Sorensen 1990; Tickner 1992; Ullman 1983; Vogler 1993; Vogler and Imber 1996.

16 A number of studies have formulated a theory of conflict based on resource scarcity (for instance, Gleick 1991, 1993a; Gurr 1985; Timberlake and Tinker 1985; Westing 1986). Food, water and oil issues present examples of scarcity playing an important precipitating role in conflict. This resource scarcity literature does not necessarily employ the term 'environmental security', but it does provide support for introducing the element of non-military threats into the modern conception of security (Dabelko and Dabelko 1995).

17 For a thought provoking discussion about the theoretical implications of applying environmental security as a universal value; the definition of the concept of environmental security in recent literature; and the consequences of an emerging global norm of environmental security for the theory of international relations, see Hugh Dyer (1996).

3
Water Scarcity: A Global Problem?

Water management: a global challenge?

Water management has been the principal challenge for humanity from the early days of civilisation, and it could be the first resource that puts a limit on human population and economic growth. Great ancient cities such as Babylon, Ur, Memphis, Athens, Carthage, Alexandria and Rome relied on their water engineers, who found the relevant water resources, brought them to their cities, arranged for storage and distribution, and made efforts at proper sewage disposal. Today, several of the major issues confronting humanity are still related to water: including pollution, desertification, waterlogging of agricultural lands, flooding, and water scarcity-related problems. Moreover, a new problem, global warming, is likely to make things much worse, because, as the earth heats up and the climate changes, the pattern of rainfall will change unpredictably – dry areas will become wet, wet ones dry, and both consumers and engineers will be hard pressed to adapt to the change. These issues have led some observers to avow that the problems of both water scarcity and water quality have reached global dimensions (Stumm 1986; Falkenmark 1990; Postel 1992; World Bank 1995). For example, Biswas confidently predicts that 'a global water crisis is imminent' so that 'by the end of this decade, water will be the critical resource problem of the world, somewhat similar to what the energy crisis was in the 1970s' (1991a: 36; 1991b: 144). How convincing is the argument that water scarcity is a global problem?

A comprehensive inventory of the water scarcity issues throughout the world is beyond the scope of this study. However, an examination of some important cases from Europe, North America,

China, the Indian subcontinent, Central Asia and the Middle East will show several common levels of concern. For example, expanding water usage is increasing transnational dependencies at a global level; the competing and conflicting demands for *'good water'*[1] is a universal problem rather than an issue confined to arid and semi-arid regions such as North Africa and the Middle East; and how water resources are used and managed locally and regionally is critically important. The findings of this chapter also reinforce the central hypothesis of the book, that complex socio-economic systems, increasing population, urbanisation and industrialisation all join to create a situation demanding rational, coordinated, long-term planning and cooperative management, rather than conflict, between riparian countries. But first, let us turn to the task of defining the key terms 'water scarcity' and 'global problem'.

Conceptualisation

What is water scarcity?

At the most basic level, the word 'scarcity' is used to indicate absolute deficiency of supply, as for example, when we say: 'Water scarcity is a constant feature of the Sahara Desert'. Sometimes it is used to imply any reduction of the supply to which people have been accustomed. In this sense we say, for instance, 'Due to very low precipitation, this year water will be scarce'. In an economic sense, we use the term scarcity when demand exceeds supply; this kind of scarcity is neither an absolute deficiency of supply nor a fall in the usual supply, but rather a situation in which more of a commodity is demanded in the market than is being supplied *at the going price*. Regarding this definition of scarcity, economists suggest that in order to solve the problem we must simply *raise the price* (see Chapter 2). There is also a sense of psychological scarcity, which exists when, for whatever reason, people believe a shortage exists and behave accordingly – whether or not physical reality justifies the impression. According to Cotgrove (1982), perceptions of the amount and quality and availability of water are usually part of a people's attitude toward the environment (see also Naff 1994b: 253–4).

If water is absolutely deficient in a region, it cannot afford extensive habitation, unless the product yielded there by human effort is sufficiently valuable to produce a surplus to fund the cost of transporting water to the region, in which case it would be possible to overcome the problem of absolute deficiency. It is also possible that, because of some non-economic considerations, settlements in such a region could be subsidised by government or by an interested group. What was done in Palestine after the Second World War by the Zionist movement, who demanded unrestricted immigration of world Jewry to an arid environment and facilitated extensive irrigation there, is a striking example of its kind. The Great Man-Made River in Libya is another example of long-distance water transportation based on non-economic considerations.

Though there are various reasons for water scarcity, they are all variations of four basic causes. First, climatic factors; second, human activities which pollute water at a rate faster than the source can be recovered; third, over-consumption which depletes a source, such as an aquifer, at a rate faster than it can be replenished; and fourth, redistribution for other uses or to other places – out-of-basin diversion. Whatever its cause, the consequences of water scarcity are immense: it has a powerful effect on the land and its people. For example, water scarcity often causes large-scale migrations from the rural areas to the cities, creating massive waves of displaced and unemployed people. Such occurrences seriously drain the economy, create social disturbances, and may lead to political instability (Gurr 1985; Postel 1992).

Is it a global problem?

What do writers such as Davey (1985), Falkenmark (1990), Biswas (1991a, b), Elliott (1991) and Gleick (1993b) mean when they talk about the 'global water crisis' or the 'global politics of water'?

The concept of a 'global problem' is used to distinguish between international issues which could be dealt with exclusively at a regional level, and issues which affect, in one way or another, the world community and demand global action for their resolution. According to Soroos (1986), there are three basic categories of policy problems that dictate a coordinated international response, the first of which could be regarded as regional in scope, and the second and third as global in scope:

1 issues which pertain to relations between states;
2 issues which affect domains outside the political jurisdiction of states, in particular global commons; and
3 domestic affairs which are common to many countries, either because it is an advantage to learn from one another's experience or to pool efforts to address the problem, or because the

outside community takes a special interest in conditions within a country. It is evident that water issues involve Soroos's first category, namely the relation between states, given the geopolitical facts that more than 47 per cent of all land falls within international river basins; nearly 50 countries on four continents have more than 75 per cent of their total land in such basins; more than 250 basins are multinational, including 57 in Africa and about 50 in Europe; and more than 40 per cent of the world's population lives in these basins (UNWC 1978; Vlachos 1990; Butler 1995). Out of the world's 214 'shared' rivers, 148 flow through two countries, 31 through three countries and the rest through four or more countries. Twelve of these latter rivers are shared by five or more countries (see Table 3.1). Not surprisingly, therefore, the security of water sources is one of the most crucial elements in foreign policy for many countries. For example, Bulgaria, Cambodia, Botswana, the Congo, Gambia, Sudan, Egypt and Syria all receive more than 75 per cent of their water from rivers flowing from upstream neighbours (Butler 1995: 34).

Turning to Soroos's second category, the global commons, today there are 5.6 billion people on earth, and by 2025 there will be 8 billion, all competing for the available water. In general, chronic water shortages affect 40 per cent of the world's population, spread across 80 nations, and the demand for water is increasing at 2.3 per cent annually – that is, doubling every 21 years (Butler 1995). Many countries, not only those in Africa and the Middle East, but also developed countries in humid regions of the earth, are now chronically short of safe water; and more are likely to join them in the future. Moreover, as Table 2.1 shows, the greater part of both fresh water and sea water is located outside state territorial boundaries. Furthermore, global environmental developments such as climate change could greatly complicate international water problems by shifting the patterns of rainfall and runoff to which agriculture and urban water systems are geared. It could be argued, therefore, that the water problem is a global issue, a tragedy of the commons on a global scale.

Finally, the water issue also seems to meet Soroos's third category in that, although the global water crisis is not likely to shake the world in the sudden way that the energy crisis of the 1970s did, nevertheless, with key crop-producing regions of the world showing signs of water scarcity and depletion (Postel 1992; Gleick

Table 3.1 International river basins with more than four riparian countries

River	Riparian countries
Danube	Switzerland, Germany, Austria, Czech Republic, Slovakia, Hungary, Croatia, Yugoslavia, Romania, Bulgaria, Moldavia, Ukraine
Nile	Burundi, Rwanda, Tanzania, Uganda, Kenya, Central African Republic, Zaïre, Ethiopia, Sudan, Egypt
Congo	Rwanda, Burundi, Angola, Central African Republic, Zambia, Tanzania, Zaïre, Cameroon, Congo
Niger	Guinea, Côte d'Ivoire, Mali, Benin, Burkina Faso, Togo, Cameroon, Niger, Nigeria
Zambezi	Angola, Zambia, Namibia, Botswana, Zimbabwe, Malawi, Tanzania, Mozambique
Rhine	Switzerland, Austria, Liechtenstein, France, Germany, the Netherlands
Amazon	Peru, Ecuador, Bolivia, Guyana, Colombia, Venezuela, Brazil
Lake Chad	Central African Republic, Cameroon, Chad, Nigeria, Niger
Volta	Mali, Burkina Faso, Côte d'Ivoire, Togo, Benin, Ghana
Mekong	China, Burma, Laos, Thailand, Cambodia, Vietnam
La Plata	Bolivia, Paraguay, Brazil, Uruguay, Argentina

Source: Compiled by authors

1993b) and with chronic water shortage in many metropolitan areas, world food supplies and economic health are significantly affected. Hence, although problems related to the predicament of arid and semi-arid areas are primarily regional in scope, as Falkenmark (1990) asserts, their far-reaching effects on the world economy, trade, and security do turn them into issues of 'common concern' for the international community. Furthermore, the transfer of effective techniques for coping with contamination, aridity, scarcity, and recurrent droughts, now being developed in some countries, is another issue of common interest. These circumstances clearly match the third of Soroos' conditions.

According to Clarke (1991), the experience in Europe implies that countries which use less than 5 per cent of their total runoff have few water management problems; those which use 10–20 per cent usually have some serious water problems; and above 20 per cent, water supply management becomes a major national issue which

Table 3.2 Water use by continent (bcm/year)[a]

	1950	1960	1970	1980	1990	2000 (projected)
Asia	855.2	1220	1520	1910	2440	3140
Africa	56	86	116	168	232	317
N. America	286	411	556	663	724	796
S. America	59	63	85	111	150	216
Europe	93.8	185	294	435	554	673
Oceania	10	17	23	29	38	47
Total	1360	1982	2594	3316	4138	5189

Sources: Meybeck et al. (1989); World Resources Institute (1990)

Note: [a] One billion cubic metres (bcm) equals one km^3 or 1×10^9 cubic metres

may be a limiting factor in economic development. If we accept Soroos's criteria and Clarke's classification and also acknowledge the pervasive problem of water pollution, it is clear that the water crisis is not restricted to developing countries, nor even to arid or semi-arid areas. The scope of water problems is world-wide, and there is an essential interdependence between countries due either to their sharing the water in the same basin, or to their getting water from the same global circulation system in which pollutants are circulated. Therefore, global problem-solving policies are undoubtedly needed to address many of the water problems plaguing the world today.

Global water demand

Excluding the polar zones, the world's runoff from the land surface is estimated to be about 40 600 km^3 per year.[2] Evenly distributed and before being polluted, this amount of water would be enough to support a population perhaps 10 times larger than the world's today (Clarke 1991). This fact raises the question: where do water problems really start? To answer this question, let us examine the world's water demand, its unparalleled increase during the last few decades, and the feasibility of ongoing efforts to develop the supply in different regions.

While there is no substitute for the supply of water, unlike most resources, the world's total water consumption has quadrupled during the last 50 years (Table 3.2). In 1940, total water use was about 1000 km^3 per year. By 1960, it had doubled and by 1990 it had doubled again (Clarke 1991: 66). Table 3.2 illustrates the trend of water consumption in each continent. This rate of increase in consumption cannot continue indefinitely, for two reasons; first, about 44 per cent of the world's reliable runoff has been made accessible through the construction of dams, reservoirs and conveyance systems (Clarke 1991: 10), but due to geographical constraints these supplies cannot be greatly increased; second, the world-wide chemical contamination of aquatic systems, which is ceaselessly destroying existing accessible supplies, threatens both land and water productivity (see Stumm 1986).

In tackling the water problem at national levels over recent decades, water resources development has proceeded at a very rapid pace. Countries have been involved in immense projects, such as the construction of transcontinental canals, to overcome the deficits that threaten to halt their development. United States, Russia (previously the USSR), and China have undertaken projects of this type. In their efforts to obtain reliable water supplies countries have, on occasions, literally moved mountains. They have dammed and flooded, piped, pumped, diverted and channelled, and altered the courses of major rivers. They have also developed the technology to extract groundwater in large volumes and from increasing depths, and to remove the salt from sea water. Furthermore, they have investigated the economics of using super tankers to deliver water like oil, and have even studied the prospect of towing icebergs from the Antarctic to melt down for tap water.

In the early days of nuclear power, when optimists believed that the time for everlasting cheap energy had finally arrived, it was thought that freshwater problems would be similarly solved by desalination plants. As the technocrats talked of such utopian ideals as 'the final conquest of disease' and the 'greening of the desert', they dreamed up plants for enormous agro-industrial complexes, in places such as Israel's Negev Desert, which would be powered by nuclear reactors and watered from desalinated sea water.

The cheap nuclear power never materialised, however, and it turns out that the business of extracting salt from sea water is actually very expensive. This means that its use is ruled out for large-scale applications such as irrigation. But desalination is useful for supplying domestic and even industrial water to those who can pay for it, mostly in arid and semi-arid regions where other supplies are very limited. Though expensive, desalinated water is often cheaper than water hauled over long sea routes by tanker.

However, despite all these technological efforts and breakthroughs, in many parts of the world, water could be the first resource that puts a limit on population and economic growth. Sandra Postel (1992: 18) claims that in several regions of the world 'water use is nearing the limits of natural systems; in some areas, those limits have already been surpassed. A number of areas could enter a period of chronic shortages during this decade, including much of Africa, northern China, pockets of India, Mexico, the Middle East, and parts of western North America.' Bearing in mind the fact that currently 36 per cent of the global harvest comes from 16 per cent of the world's irrigated cropland, and that 84 per cent of the world's cropland which is watered only by rain is exposed to global warming and climatic upheaval (see Lean 1995a, d), the logical outcome of Postel's observation is that global shortfalls of water will mean global shortfalls of food.

Regional water problems

It is evident from the above discussion that a global water crisis is approaching. Assessing the extent and severity of the crisis in different parts of the world demands a detailed analysis of their climatological and hydrological characteristics, as well as the socio-economic achievements and capabilities of their inhabitants. However, since such an analysis is well beyond the limits of this study, in the following pages we shall try to illustrate the extent and gravity of the water crisis in some selected areas in Europe, Southwest America, Mexico, China, India, Central Asia, and the Middle East.

Europe

People living in industrialised temperate regions generally take water for granted. This perception, as Falkenmark (1990) rightly observed, has caused blindness to water scarcity – people have failed to appreciate the fact that water is a finite resource. Pollution reduces supplies much more than drought, as each gallon of dirty water contaminates many more gallons in the receiving river or lake. It is estimated that, in general, for every unit of waste water which is allowed to flow into rivers and streams, a further 13 units of water are needed to transport and dilute this contaminated water before it can be used again (Clarke 1991: 24). Seen in this light, water availability appears even more problematic than it seems at first glance.

Europe, with its temperate climate and highly developed river basin network,[3] suffers from major pollution problems, including water pollution. In this humid continent, water resources are fouled by sewage, toxic chemicals and acid rain. For instance, the Danube, the longest and the most important river in Central Europe and the continent's second-longest river after the Volga, is endangered at every turning by runoff from nitrogen-rich agricultural fertilisers and by the industrial plants that discharge synthetic chemicals, mercury, cadmium, lead, zinc, nickel and arsenic along its banks, from Switzerland, where it rises, to Romania, where it pours into the Black Sea. It is so contaminated that some observers call it a 'deadly swamp' (Painton 1990: 40). In the Rhine River basin, the longest river in Western Europe, an analysis of metal concentrations revealed that, for example, cadmium concentrations in sediments had risen from less than 2mg/kg before 1920 to about 40mg/kg in the early 1970s (Nash 1993). As a result of pollution, more than 20 million people in Rhine's riparian countries, particularly in the Netherlands, must have their drinking water filtered to remove heavy metals, hazardous chemicals, and salt deposited further upstream in France, Switzerland and Germany (Renner 1989b). Moreover, sometimes massive chemical spills severely affect the water system of the riparian nations.[4]

In general, fewer than 10 per cent of the rivers in Europe can be classified as pristine (Meybeck *et al.* 1989). However, while Europeans recognise their pollution peril, the effort to clean up their environment clashes with their desire to boost consumption of food and manufactured products, and to avoid unemployment. This situation has affected relations between domestic authorities as well as relations between neighbouring countries. In the UK, for example, there was a dispute over charge increases between a group of about 30 textile companies, on the one hand, and local water authorities, on the other hand (Hotten 1995). The textile companies were arguing that increased bills were likely to affect their ability to export and thus result in big job losses, while the water authorities blamed the textile firms for some of the worst pollution and claimed that the price rises were necessary to deal with the effluent treatment.[5] At the same time, the Department of the Environment published figures which showed that Britain has been among the worst performers in reducing its discharges to the North Sea over the last 10 years (Nicholson-Lord 1995).

In Poland, the share of river water of the highest quality for drinking has dropped from 32 per cent to less than 5 per cent during the last two decades. Some three-quarters of that nation's river water is now too contaminated even for industrial use and 95 per cent of it is unfit to drink. The Vistula River, which runs through Poland, is so laden with poisons and corrosive chemicals that when it reaches the Baltic Sea, it pushes a plume of polluted water 150 miles to the Swedish coast (Lean 1993; Postel 1992). The degree of water pollution is such that Painton (1990: 40) declares that 'much of the food grown in Silesia is so contaminated by toxins that it is unsafe to eat'.

While the situation in Eastern Europe is much worse than in the west, the median level of some chemical concentrations in the rivers of this water-rich continent is much higher than elsewhere in the world. The World Health Organization's Global Environmental Monitoring System has observed (Meybeck *et al.* 1989), for example, that the median level of nitrate concentration in the rivers outside Europe is 0.25 mg/l, while for Europe it is 4.5 mg/l. As Linda Nash (1993: 37) points out, the 'uncontrolled use of chemicals is putting ever greater stress on the natural environment, and the depletion and degradation of water resources are causing the costs of new water supplies to rise'. This is because even the more developed countries have failed to address the question of source reduction and comprehensive watershed management, and have instead relied on much more costly monitoring and treatment strategies, such as the use of expensive water-purifying equipment (Aczel 1991). Treatment of contaminated water is only part of the answer; some measures should be taken to reduce the contamination of water sources so that treatment becomes less necessary. Thus, as Nash (1993: 37) has rightly shown, without new approaches to both development and environmental protection even in temperate regions 'the widespread degradation of water quality that we currently face could become an unmanageable crisis'.

America

The USA

Throughout most of US history, government actions have encouraged human exploitation of natural resources: logging, mining, drilling, grazing, damming rivers.[6] This exploitative philosophy reached its height during the Reagan years. Some Americans have recognized that 'America's ecology is in crisis' (Gup 1993: 38), not least because, for all its natural wealth, the US has its share of water woes. The water Americans drink may look clear and clean, but it often contains noxious chemicals[7] and malicious microbes. It is true that overall, the US still has one of the cleanest water supplies in the world, but that does not mean it is safe in all places at all times.[8] Nearly half of its rivers, lakes and streams are damaged or threatened by pollution, according to a survey conducted by the US Environmental Protection Agency (EPA) (*TIME*, 20 August 1990: 61).

Drought is another problem. In 1989, a drought baked the US soil from California to Georgia, reducing the country's grain harvest by 31 per cent and killing thousands of head of livestock (Sancton 1989). Occasional water shortages occur all over the country, even in the rain-rich Northeast. By far the most serious problems, however, are in the West, where poor water-management practices and extreme drought have been detrimental to both population and vital ecosystems. In cities and towns, water scarcity means quick showers, brown lawns and dirty cars. But the real economic burden falls on farmers, who use between 80 per cent and 90 per cent of the water available in the Far West.

Former Governor Bruce Babbitt of Arizona was quoted as saying that: 'Only very recently has it become clear that there are no more water holes in the West' (*Time*, 20 August 1990: 61). Even with conservation, therefore, some parts of the West will remain vulnerable. Southern California gets about half of its water from a single canal system, the California Aqueduct, which carries water from the Sacramento River Delta 800 km south to Los Angeles. The area is also the hub of a huge network of dams, canals and pumps that divert water to irrigate the Central Valley and supply 19 million users in the semi-desert southland. However, the price of this development, Marc Reisner (1990) claims, has been a series of ecological calamities. For instance, the delta is sinking by as much as 7.6 cm a year, leaving the area, much of it already below sea level, ever more vulnerable to seawater intrusion. A major earthquake could destroy the levees that protect this system. Another river in the region, the San Joaquin River, is entirely diverted for irrigation as it emerges from the Sierra Nevada, and when it resumes downstream near the Kesterson Reservoir, selenium-poisoned waters flow into it from the Westlands agricultural district (Linden 1991: 83).

While desert covers one-fourth of the state, each year 50 000 acres of California's cropland give way to housing tracts or shopping malls. The consequences of these developments are so dire that California's resources secretary has been quoted (Linden 1991: 83) as predicting that the time will come when large companies begin to flee California because of ecological as well as other problems; 'The point at which a major company gets fed up with bad air, scarce water, housing prices and traffic, and talks about future capital spending in Colorado or Arizona is the point at which you get a political response.'

Environmental disasters in the USA include a 12-feet fall in the level of Utah's Great Salt Lake; the depletion of the aquifers that supply groundwater to eight Great Plains states (Elmer-Dewitt 1989); losing land to the sea; and the intrusion of salt water into the depleting aquifers. For example, in the US Midwest, the so-called 'bread basket' of the world that largely depends on the giant Ogallala aquifer which stretches for 1000 miles across eight states, so much is being taken from this hidden sea that its level is dropping by four feet a year. If this continues, some observers believe the agricultural abundance of the Great Plains will last only another 20 years (Lean 1995d). A primary study by the EPA suggested that the cost of protecting US developed coastal areas could reach $111 billion (Elmer-Dewitt 1989). Southern Louisiana, which is losing land to the Gulf of Mexico at the alarming rate of one acre every 16 minutes, has already initiated ambitious programmes, including a $24 million pumping station which would divert millions of gallons of silt-rich Mississippi River water onto the coastline to help stop saltwater intrusion and to supply sediment that will build up the eroding land. However, as giant pumps divert fresh water into canals, the zone where fresh water and salt water meet moves upstream and affects the whole basin and its life-giving system.

Facing such conditions of scarcity, North Americans have proposed a new version of transferring northern waters toward their arid lands. The $100 bn megaproject named the GRAND (Great Recycling and Northern Development) Canal has been suggested to create a new fresh water reservoir by blocking off Canada's James Bay and channelling its water to the arid west. According to another proposal, the North American Water and Power Alliance (NAWAPA) would transfer water from rivers in Alaska and Western Canada[9] to the American Southwest as far down as Mexico (Postel 1992: 43). The feasibility and suitability of such projects, however, is under question. In spite of their traditional policy which has encouraged human exploitation of natural resources, Americans are learning that there are limits to mankind's ability to move water from one place to another without seriously upsetting the balance of nature.

Mexico

In Latin America, one of the most troubled dry spots is Mexico, where more than 30 million people do not have safe drinking water. A government report asserts that 'water will be a limiting factor for the country's future development' (Linden 1990: 58). While the nation's capital suffers from chronic water shortages, the demands of over 20 million people are causing the level of their main aquifer to drop as much as 3.4 metres annually. Says Ivan Restrepo, head of the Centre for Ecodevelopment in Mexico, 'We've been enduring a crisis for several years now, but it is in this decade that it will explode' (quoted in ibid.: 58).

Northern Mexico is dependent on two major rivers, the Colorado and the Rio Grande. Within the last decade Mexico has continuously suffered from the increasing concentrations of salt in the Colorado River. Some 80 per cent of the Colorado delta's arable land, which contains 200 000 hectares of the most productive land in Northern Mexico, has already been affected by salinisation, and 14 per cent is said to be too saline for cultivation. This process, according to some American calculations, is going to end agriculture in the Mexicali Valley by the year 2000 (Timberlake and Tinker 1984). The main causes are an irrigation project in Arizona which drains highly salty water into the Colorado, not far north of the border, and a large dam at Glen Canyon in the USA which reduces the freshwater flow downstream. In the Rio Grande, which forms much of the boundary between the two states, the rate of extraction is so high that the river is almost dry in summer (Calvert 1993). However, the USA uses a 1906 agreement on this river to justify unilateral reductions in water supply to Mexico during years of serious drought, as in 1978. A similar problem faces the Juarez aquifer system which is shared by the two countries. Since the turn of the century the aquifer's water table has steadily dropped[10] and its natural recharge has failed to maintain the levels required. Mexico claims that uncontrolled proliferation of wells on the Texan side is jeopardising the water sources of Mexican towns near the aquifer. The threat of growing water scarcity has resulted in growing tension between the two nations (Timberlake and Tinker 1984).

Asia

China

China, which is endowed with 8 per cent of the earth's freshwater resources, also faces serious water constraints because it has to feed

22 per cent of the world's population (Postel 1992: 34). Although China's experience in water resources development is one of the oldest in the world,[11] during the past 40 years its water resources development projects have reached unprecedented magnitude. For instance, the number of large dams built in China during this period exceeds that of the rest of the world (Le Moigne 1994a: 7). However, according to officials from China's Ministry of Water Resources, serious water shortage in some regions has become one of the main factors constraining the development of industry and agriculture (Sun 1994: vii).

China's seven main river basins together account for 86 per cent of the country's population and 51 per cent of its area. Some 40 per cent of China's population and 60 per cent of the gross national production of industry and agriculture are located on the middle and lower reaches of these seven rivers (Sun 1994). Being well aware of the important role of water resources in the national economy, from the 1950s onwards, the Chinese government focused on comprehensive water resources development and farmland water conservancy. Due to investment of approximately $50 billion, an enormous financial burden of well over 2 per cent of China's GNP, by the end of 1988, some 80 000 reservoirs of different sizes had been built, with a total storage capacity of 450 billion cubic metres (bcm), and 175 000 kilometres of river dikes were constructed, as well as numerous irrigation canals. This process, it is widely believed, has acted as a strong catalyst for national economic growth. The irrigated area, for example, increased from 16 to almost 50 million hectares, accounting for 47 per cent of China's total cultivated land and 20 per cent of the world's total (Wu 1994). However, during this period, as a result of these technological advances in water supply engineering, nature's equilibrium has been upset. Some 10 million people have been displaced (Sun 1994), and the natural environment has severely deteriorated, while water still stands as a significant limiting factor for the country's future development.

As China entered the 1980s, the issue of environmental protection gradually achieved an important place in the nation's economic and social activities. The State Environmental Protection Committee was set up and the Environmental Protection Law was passed. As a result, water quality monitoring and control over waste discharges have been strengthened, waste discharge fees and pollution fines are being collected, while the use of technological processes has reduced the volume of waste water discharged. However, water pollution remains a serious problem. For instance, in 1985, the volume of polluted water discharged into various bodies of water was 34.2 billion tons, exacerbating the already serious shortage of water. By the year 2000, Bingcai (1994) notes, this figure is expected to increase to 90 billion tons.

Most rivers, lakes, and aquifers have been polluted to various extents. In some regions, so much ground water has been extracted that some aquifers have been pumped dry and ground settlement has become obvious (Quingquan 1994). In Beijing, for example, one-third of the city's wells have run dry, and the water table drops by as much as 2 metres per year (Linden 1990; Postel 1992: 35). In some places the water is unsuitable for drinking and, during dry years, water is supplied only at certain hours. Indeed, about 200 cities are suffering from water shortages.

Considering China's huge population and its continuing growth, and taking into account the fact that 40 per cent of the country's population and 60 per cent of its industry and agriculture are concentrated along the rivers and streams (Bingcai 1994), it is quite evident that, as the economy develops further, the annual discharges of waste water and polluted water will steadily increase. The predicament is particularly severe in and around Beijing, the important industrial city of Tianjin, and the North China Plain, a vast expanse of fertile farmland that yields a quarter of the country's grain. Planners have predicted that Beijing's total water demand in the year 2000 will outstrip currently available supplies by 70 per cent. For Tianjin, optimistic projections indicate a shortfall of 36 per cent (Postel 1992: 35). These circumstances threaten to cause critical consequences for future generations.

China's policy-makers are fully aware of the crisis. According to Wu (1994), 'the discrepancy between the demand for water and the supply available for industry, agriculture, and the urban population becomes more pronounced with each passing day'. As demands continue to rise, water supply projects get more expensive and difficult to build and, thus, water budgets are becoming badly imbalanced. Chinese leaders, therefore, have at last begun to focus the nation's scientific talent on the task of developing salt-tolerant and drought-resistant crops (Linden 1990: 83).

India

The sad state of India's water supply is just one sign of what could become a global disaster. In large cities such as Madras, India's fourth-largest city, there is tap water only for a few hours per day.

In rural areas, at least 8000 Indian villages have no local water supply and their residents must travel long distances to obtain water. 'Tens of thousands of villages', declares Postel (1992: 36), 'across the subcontinent experience shortages, and their numbers are growing'. Besides, in many parts of the country, water is contaminated by sewage and industrial waste, exposing those who drink it to disease (Linden 1990).

For a country like India, which gets 80 per cent of its rainfall in three to four months, with much of it coming in just a few monsoon storms, the ability to control water and moderate its release means the difference between devastating floods and droughts and a manageable year-round supply. While there is often too much water during the monsoon season, there is often too little during the subsequent dry season. Bearing in mind the fact that half of India's territory is arable and almost 25 per cent of it is irrigated (Grolier 1993), it is quite obvious that the survival of India's national economy requires a sound environmental base capable of providing a steady supply of fresh water.

Indeed, India's success in turning itself from a severely food-deficient nation into a nation with a slight food surplus, according to Myers (1993), is due mainly to its irrigated agriculture, because productivity of the rain-fed farmlands is very poor. India's irrigation water supplies are critically dependent on forested catchments which are designed to absorb rain water and moderately release it through an intricate network of irrigation canals. Unfortunately, however, deforestation has led to severe disruption of these critical watershed systems.

Though excessive fuelwood gathering is the principal cause of India's deforestation, ironically, it is partly due to major irrigation projects, such as dams. For example, the controversial reservoir of the Sardar Sarovar Dam alone would flood 37 000 hectares of forest and farmland.[12] As far as demand for fuelwood is concerned, in 1982 it was 3.5 times greater than the amount that the Indian forests could sustainably supply. By the end of the century, the gap will be much larger as India's population continues to grow and its forests continue to contract; even if all planned tree plantations are established, Myers (1993) predicts that there will be a shortfall twice as large as that of 1982.

The collapse of watershed functions has combined with rapidly rising demand, due to population growth, irrigation expansion, and industrial development, to compound pressures on India's water resources. Groundwater overdrafting is also now widespread in several parts of India, including the states of Andhra Pradesh, Karnataka, Maharashtra, and Tamil Nadu. Reports indicate, for instance, that heavy pumping in portions of the Southern Indian state of Tamil Nadu lowered the water table by as much as 25–30 metres in a decade. In the Western state of Gujarat, overpumping by irrigators in the coastal districts has caused salt water to invade the regional aquifer, destroying existing freshwater supplies (Postel 1992: 54).

Various parts of the country face severe chronic or seasonal water shortages, and federal and local governments frequently clash over water resources allocation. This predicament recently led to conflict in the Punjab State, where Sikhs claim that too much of 'their' water is being diverted to the Hindu states of Harayan and Rajasthan. These clashes contributed to the military storming of the Sikh Golden Temple in Amritsar in 1984 and, consequently, to the assassination of the prime minister, Indira Gandhi (Timberlake and Tinker 1985; Clarke 1991; Grolier 1993). Moreover, Punjab lies next to the border with Pakistan, an area susceptible to tension over the division of water flows from the Indus River system (see Yunus Khan 1990; Kirmani 1990). In the present situation, one-third of India's population suffers from malnutrition and much more from lack of safe water and sanitation. The question India faces is what will happen by the turn of the next century when, as some observers maintain (see Myers 1993; Hassan 1991), its irrigation demand is projected to increase by 50 per cent, industry's demand to be at least twice as high, and population well over 1 billion?

Central Asia

Signs of water-borne environmental disasters are also evident in Central Asia. By far the most dramatic is the shrinking Aral Sea, once the fourth-largest inland lake in the world. Mismanagement of land and water around the Aral Sea has cut it off from its sources of water, the Syr Darya and Amu Darya, causing its volume to drop by nearly 70 per cent between 1960 and 1990. Consequently, its area has shrunk about 50 per cent, and its salinity has increased by about threefold – a trend which is still continuing (Golubev 1990; Engelmann 1994; Kirmani and Rangeley 1994). There is no controversy about the cause, says Clarke (1991); the root of the problem is over-irrigation. The two rivers which drain into the Aral Sea, with a combined flow of 100 bcm/year (Shabad 1986), have been heavily tapped to irrigate a semi-arid desert where nearly all agriculture relies on irrigation. Expanding the area under irrigation up to 7.5 million hectares (Encarta 1994) has meant diverting more and more water from the rivers that supply the Aral Sea into an extensive canal system. Karakum canal, for instance, diverts annually about 12 km^3 of water 800 kilometres away from the Amu Darya.[13] Moreover, as Clarke (1991) declares, the Syr Darya has been so heavily tapped for irrigation that its water never reached the Aral at all between 1974 and 1986; and the Amu Darya failed five times between 1982 and 1989 to do so.

The shrinking of the Aral Sea has caused ecological and environmental calamities. The lower reaches of the rivers once contained large deltas that supported an extensive vegetative cover, but they have disappeared due to reduced water flow (Engelmann 1994). Because of ever increasing salinity, once a fruitful sea, now Aral has turned into a poisoned wasteland. Virtually all of its native fish species have disappeared leading to the collapse of the region's fishing industry. According to Clarke (1991), the fish catch, once 25 000 tonnes a year, has been reduced to zero. Thus, once-populated shores are now barren and uninhabited. Several villages and large towns, such as Aralsk and Muynak, which were active ports before 1960, now lie 48 kilometres inland (Grolier 1993) and as a consequence of economic ruin, masses of people have left the cities – for example, the city of Aralsk has lost half of its population (Clarke 1991).

Moreover, as the lake has dried up, various types of salts have precipitated on its former bed; and since most of these salts are subject to wind erosion, the exposed bed of the Aral has become the main source of salt and dust storms which contaminate the land and, as Linden (1990) reports 'afflict millions of people with gastritis, typhoid and throat cancer'. The amount of damaging salt which wind dumps annually on surrounding cropland is well over 40 million tons, according to Micklin (1989). Besides salinisation, waterlogging is another major problem in Aral's basin, because two-thirds of the irrigation canals are unlined (Shabad 1986). Moreover, since many of the irrigated lands around the canals are at a lower elevation than the Aral Sea, returning used irrigation water to the Aral is prohibitively expensive (Engelmann 1994).

Finally, these disastrous water policies have compounded the tragedy by sinking the historical heritage of the region's ancient civilisation. The priceless archaeological and architectural landmarks of the region have collapsed or are falling to pieces. Hundreds of buildings in Samarkand, Khiva, and Bukhara, which according to Specter (1995), are considered to be among the most exotic places on earth, are unlikely to survive into the next millennium. The United Nations in 1994 declared that Bukhara and Khiva are among the architecturally threatened centres of ancient civilisation; and some international experts were dispatched to help find out practical ways for halting the decay created by the excessive waterlogging of the region. However, continuity of those policies that have run rivers dry and have left water accumulating where there should not be any at all, means that every day some parts of this precious and delicate heritage disappear. 'In the 1960s there were 12,000 important archaeological monuments in this country,' said Shirinov Temyr, the director of the Uzbek Academy of Science's archaeological division. 'Since then we have lost at least 5,000 of them' (Specter 1995).

Unfortunately, these trends are continuing. Golubev (1990) predicts that by the turn of the century, the sea level will have dropped another 10 metres or so, while its salinity will have tripled again. Continued unsustainable development of water resources in this region is going to create an even worse crisis in the future – exacerbating ecological, demographic, economic, and cultural problems. 'Reviving the Aral Sea may not be possible at all', says Gleick (1993b). And it seems unlikely to Clarke (1991) that 'the 600 cubic kilometres of water that the Sea has lost can ever be put back'. However, preventing its further drainage is now a high priority in the regional countries. But saving the Aral cannot be done without a complete change in the style and form of water use and management in the region. Officials and scientists who have studied the problem say that restoring the ecosystem will require major reductions in agricultural water use. This could be achieved by improving irrigation efficiency; or by switching to less water-intensive crops; or by taking some of Central Asia's farmland out of irrigation. However, regarding the financial and economic difficulties of the regional countries, the first alternative – of improving the efficiency of such a huge irrigation system – is not affordable. It calls for both financial and technical support from the international community.

The second alternative – switching to less water-intensive crops – is difficult and time consuming because the Central Asian republics are still economically dependent on Russia and practically obliged to cultivate cotton, the most water-intensive agricultural product, in order to provide Russia's needs. Although an effort is now being made to diversify the economy, cotton will remain an important

Table 3.3 The share of agriculture in national economy and employment in Central Asian republics

Country	Net material product (NMP) (%)	Employment (%)
Kazakhstan	30+	17
Kyrgystan	28.4 (of GDP)	30
Tajikistan	40	30+
Turkmenistan	40+	40+
Uzbekistan	43	29

Source: Encarta (1994)

source of hard currency. Moreover, it is also used to barter with other former Soviet republics to obtain essential supplies. The third alternative – taking farmland out of irrigation – seems the most effective and short-term answer to the predicament of the region. But taking land out of production in one of the most productive agricultural areas, is hardly appropriate to nations for whom agriculture remains the mainstay of the economy (see Table 3.3).

It is clear that the Aral Sea and its tributary rivers – Amu Darya and Syr Darya – present one of the most challenging examples of a regional water-borne crisis which calls for international cooperation. The issue not only relates to the destruction of the Aral Sea and the river deltas, and the staggering problems of salinity, pollution, and waterlogging in the irrigated areas of over 7 million hectares which are threatening the economic stability and architectural and cultural heritage of riparian states, but also to the ability of the riparian states to make cooperative efforts to address these problems. After independence, the riparian states signed a water agreement and established regional institutions for coordinated management of their common water resources, and they have been promised support for their efforts by the World Bank, UNEP and the UNDP. Their success depends on their continued cooperation and commitment to address the formidable problems of the Aral Sea Basin.

The Middle East

In those societies located in arid environments where indigenous water is limited, the challenges are formidable. While humid regions have been endowed with a surplus of water, Middle Eastern countries are afflicted with a chronic shortage of it. The region generally faces a severe water problem, with the gravity of the situation varying among different countries. Only Lebanon, Turkey, and Iran among Middle Eastern countries have adequate *indigenous* water resources to meet their current and future water needs, including those of agriculture. According to the World Bank, the costs of water supply and sanitation in the Middle East are the highest in the world, being over twice those in North America and five times those in Southeast Asia (Starr 1991: 18). However, some observers believe that the Middle Eastern countries now need twice as much water as is currently available, and by 2025 they will probably need four times as much as is available in their indigenous natural resources (Allan *et al.* 1995: 1).

Historically, the water shortages besetting the Middle East, though endemic and formidable, have never been insoluble. Through complicated series of dikes, canals, reservoirs and lifting devices, the early inhabitants of the region managed to transform the 'hydrological chaos' of the great river valleys into the well-organised and regularly watered fields, meadows and gardens which in the Euphrates–Tigris valley were known mythologically as the Garden of Eden (Clarke 1991: 14). However, a disturbing trend has recently emerged: over the last few decades, the average available supply of water for the entire Middle East has fallen rapidly. Presently, more than half of the Middle Eastern countries are confronting serious water shortages (see Table 3.4). Water scarcity is increasing year by year due to persistent population growth, over-exploitation, and pollution of existing resources.

The Middle East water crisis stems primarily from its climate – wide seasonal temperature variations and highly irregular rainfall. On the whole, the region simply does not receive enough rainfall to support even subsistence agriculture without extensive irrigation. Moreover, the effects of climate, poor supply, maldistribution and escalating populations have resulted in extreme variations of water supply per person across the region, ranging from a per capita supply of less than 10 m^3 in Kuwait to as much as 5500 m^3 in Iraq in rainy years (see Table 3.4). These figures indicate that the water crisis in the Middle East is not only a matter of absolute scarcity, but also of uneven resource distribution. Furthermore, the predominance of fossil ground water, which is the most striking hydrogeological phenomenon occurring on a regional scale in the Middle East,[14] indicates that every nation in the region is linked to another by a common aquifer subject to over-withdrawal and contamination.

In the light of the geopolitical fact that all major water resources in the Middle East – both rivers and aquifers – are shared by several countries, the water bodies in the region are classified as

Table 3.4 Per capita water availability and the ratio of its demand to supply in the Middle East

Country	Per capita water availability, 1990 (m^3 per person per year)	Projected per capita water availability, 2025 (m^3 per person per year)	Water withdrawals as % of renewable supplies
Qatar	60	20	174
United Arab Emirates	190	110	140
Yemen	240	80	135
Jordan	260	80	110
Israel	470	310	110
Saudi Arabia	160	50	106
Kuwait	<10	<10	>100
Bahrain	<10	<10	>100
Egypt	1070	620	97
Iraq	5500	n.a	43
Iran	2080	960	39
Oman	1330	470	22
Lebanon	1600	960	16
Syria	610	n.a.	9
Turkey	3520	n.a.	8

Sources: World Resources Institute (1990); World Bank (1992); Gleick (1993c)

international water resources. Where the catchment areas of water systems coincide with disputed land, the transboundary nature of the scarce water makes its utilisation even more competitive and complicated. Israel, for example, receives more than half of its water resources from occupied territories (Libiszewski 1995). This means that territorial and hydropolitical interests are inextricably intertwined in the Arab–Israeli conflict (see Dolatyar 1995). Within the context of this historical struggle, hydrological matters inevitably contribute an additional dimension to the already complex Arab–Israeli relations; a dimension of which the relative importance has been growing over recent years. This is because, according to Starr (1993: 1264–7), unless dramatic steps are taken now, by the year 2010, Israel, Jordan, and the West Bank and Gaza will have sufficient water only for drinking and industrial purposes. There will be no water for agriculture. Such a gloomy prediction suggests that, although a solution to the water crisis may not be a sufficient condition to ensure lasting peace and prosperity in the region, it is nevertheless a necessary one.

Conclusion

As stated at the beginning of this chapter, the first aim of the discussion was to show that several issues facing humanity are today related to water pollution and/ or water scarcity. The second aim was to show that the problem of water scarcity is world-wide rather than a problem confined to some particular regions. It is now clear that the water issue is a 'world-wide problem' and, therefore, calls for 'international problem-solving policies'. It is also evident that ensuring the supply of water at an affordable cost and of acceptable quality for domestic consumption, food production, and other uses is not an easy task and is going to be one of the most challenging issues of the next century. The problems of water scarcity, as several experts such as Falkenmark (1990), Clarke (1991), Postel (1992), Gleick (1993b) and Sun (1994) affirm, are common to both industrial and developing countries. Ever increasing population and its consequences have so extended the pressure on the limited water supplies that one can hardly find a country which is not, in one way or another, at the edge, if not in the middle, of a water crisis. In addition to the appalling consequences of the crisis for humanity, water scarcity has a devastating effect on the ecosystem; countless creatures are poisoned and crowded out as industry, agriculture and municipalities reroute rivers, dry up wetlands, dump waste and disrupt the normal functioning of delicate ecosystems. Moreover, it is now recognised that water is one of the two or three major factors which most constrain the socio-economic development of the international community.

Having formerly assumed that water shortages merely reflect temporary problems of distribution and, that, thanks to technology, titanic projects would solve any water-related problem, both industrial and developing nations are now realising that not only is the world's fresh water a finite and vulnerable resource, but also that there are limits to humankind's ability to move water from one place to another without seriously upsetting the balance of nature, whose capacity to sustain life is already under stress. Moreover, faced with newly approaching phenomena such as 'global warming' (see Lean 1995a,b,c; Schoon 1995) and their subsequent environmental consequences, mainly the dramatic changes in precipitation and evaporation patterns, the international community now recognises that it is going to face an even more severe challenge for survival. Clearly, since both the causes and effects of the challenge are international, they have to be tackled internationally. With this international perspective in mind, we must now turn to the main focus of this book – the problems of water conflict in the most water scarce region of the world – the Middle East. We begin with an examination of the most tense part of that region, the Jordan River basin.

Notes

1 Consideration of quality is important in evaluating water resources, because water cannot be applied to a given use unless its quality is suitable for that use. Here we are talking about water which meets the basic standards of quality for appropriate use.
2 Global runoff is assessed at 40 673 km^3 by the Institute of Geography, National Academy of Sciences of the former Soviet Union, and published in World Resources 1992.
3 It has four river basins shared by four or more countries, governed by 175 treaties (Timberlake and Tinker 1984).
4 For example, in 1986, nearly 30 tons of toxic waste, including fungicides and mercury, entered the Rhine. The spill, called the greatest non-nuclear disaster in Europe in a decade, forced the closing of water systems in West Germany, France, and the Netherlands (Encarta 1994).
5 During the 1990s in the UK, water companies are required to spend £28 billion to maintain and improve water quality standards. That is equivalent to more than £7 million a day up to the year 2000 *(European Water Pollution Control,* 2 (3): 5).
6 There are now over 230 dams on 59 river basins *(TIME,* 12 July 1993, p. 38).
7 For example, EPA figures show that about 50 million Americans drink radon-tainted water. This radioactive tasteless, odourless gas, which seeps into water naturally from underground rocks in many areas, is a proven cause of both lung and rectal cancer *(TIME,* 15 November 1993, p. 85).
8 Hardly a week goes by, in fact, without reports about contaminated water somewhere in the nation, and the incidents that make news are only a tiny part of the problem. According to a study by the Natural Resources Defence Council, there were some 250 000 violations of the federal Safe Drinking Water Act in 1991 and 1992 alone, affecting more than 120 million people. Americans are ingesting such noxious pollutants as bacteria, viruses, lead, gasoline, radioactive gases and carcinogenic industrial compounds. 'Like so many other problems that we have swept under the rug during the past decade and more,' said David Ozonoff of Boston University's School of Public Health, 'the national task of assuring that our drinking water is safe to drink can no longer be postponed' *(TIME,* 15 November 1993, p. 85).
9 Even in water-rich Canada, which has 9 per cent of the world's fresh water, there are local water shortages and widespread contamination of both surface and ground water (Science Council of Canada 1988).
10 Between 1903 and 1976, the aquifer's water table fell by 20–5 metres (Timberlake and Tinker 1984).

84 *Water Politics in the Middle East*

11 The struggle to tame China's mighty rivers was a central preoccupation of the 'Middle Kingdom' emperors for more than 4000 years. That is why some observers claim that the Chinese were the pioneers of hydraulic engineering. Testimony to the extended history of China's water resources engineering is demonstrated, for example, by the ancient and elaborate flood control and irrigation systems along the Yellow River, and the 1794 km long Beijing–Hangzhou Grand Canal, which is still the longest such structure in the world (Le Moigne 1994: 7).
12 The Narmada Valley Development Programme in western India is one of the largest water development ventures under way in the world. It encompasses 30 large, 135 medium-sized, and 3000 small dams. The Sardar Sarovar dam is the scheme's centrepiece.
13 The largest contributor to the decline in the Amu Darya's water and, as a result, to the desiccation of the Aral Sea, is the Kara Kum Canal, the longest canal in the former Soviet Union and one of the longest in the world. This canal extends nearly 1400 km from its headworks on the Amu Darya River to the town of Krasnovodsk near the Caspian Sea and carries about one-ninth of all the water diverted in the Aral Sea basin. The canal is used primarily for irrigation, while it is an important source of drinking water (Engelmann 1994).
14 It is estimated that 65 000 km^3 of ground water are stored in Middle Eastern artesian basins (Burdon 1982).

4
Water Politics in the Jordan River Basin

Introduction

While the water crisis is an expanding world-wide problem (see Chapter 3), many studies and reports tend to highlight the prominence of the issue in the Middle East (see in Chapter 1, notes 3–7). And among all the water basins of the Middle East, the Jordan River basin has been the focus of the most intensive attention. Despite its historical prominence, the Jordan River is the smallest watershed in the Middle East that is shared by more than two countries. From a hydrological point of view, the physical dimension of this river in no way matches its biblical fame. As its name in Arabic, *Nahr al-Urdunn*, indicates, the Jordan is rather a rivulet than a river in the proper sense of the word. From its furthest headwaters in Lebanon to the Dead Sea, with all the meanderings it makes in its course, the river measures about 320 kilometres. All tributaries considered, the basin drains an area of about 18 000 km². Total average intact flow of the Jordan River is between 1200 and 1850 million m³ per year. This is less than 2 per cent of the Nile, 5 per cent of the Euphrates, and slightly more than 3 per cent of the Tigris river flow (see Table 4.1). Furthermore, this already limited amount of water is subject to extreme seasonal and annual fluctuation. According to Lowi (1993a: 28), whereas in February the river may carry almost 40 per cent of its total annual flow, in the summer and autumn months, when water is most needed, it carries only 3–4 per cent of its annual discharge. Moreover, in frequent drought periods like 1987–91, the annual discharge of the Jordan basin can be reduced by up to 40 per cent (Kliot 1994: 178).

86 Water Politics in the Middle East

Despite its small size and its limited water, the Jordan River basin has been the locus of intense international conflict during the last five decades. It is also the most frequently cited case among all the water systems in the Middle East as a source of serious conflicts and the likeliest of all to create a new water war (Cooley 1984: 3; Gowers and Walker 1989: 7; Bulloch and Darwish 1993: 36; Gleick 1994a: 8; Mastrull 1995: 36).

In this chapter we shall show that, for four reasons, this conflict-orientated picture of the Jordan River basin is distorted and seriously misleading. First, it is one-sided because it focuses narrowly on the role of water in international relations, attributing a degree of strategic prominence to water which it does not warrant. Second, it ignores the fact that the water crisis in the Jordan River basin relates primarily to the mode of water allocation and use *within* the riparian states rather than *between* them. Third, it turns a blind eye to the efforts which have been made, at national, regional and international levels, to avoid water conflict in the basin. Indeed, given the fact that the Middle East has been subject to manipulation by the world powers, and given the general atmosphere of tension in the region, it is remarkable that open conflict between states over water has so rarely occurred in such a water-poor and politically volatile region as the Jordan River basin. Fourth, the water conflict that has occurred in the Jordan basin was not caused by water scarcity, but by Arab–Israeli conflict. By an analysis of the relationship between Zionism[1] and the environment, we shall show that the radical environmental disturbance in the Jordan River basin was a consequence of both politically induced massive immigration to the region and the Zionists' instrumental approach to the environment, and this has been the underlying factor in the process of water crisis in this basin. In developing this argument, we shall establish that water conflict in the Jordan River basin is a function of the Zionists' exploitative approach to the fragile natural environment of the basin. Their imperious attitude to the environment, we argue, is reflected in their political, economic, and military policies and strategies. It is concluded that, if the Israelis were to come to terms with their natural environment, the problem of water conflict would fade considerably, and that, despite some sensational political rhetoric and alarmist predictions, this process seems to be the most likely scenario in the region.

In short, by an analytical review of the evolution of the water scarcity issue in the Jordan River basin, we shall demonstrate that

Table 4.1 Comparative hydrology of the major rivers in the Middle East

	Length (km)	Total discharge (bcm/yr)	Drainage area (km^2)
Euphrates	2 320–30[a, k]	31.8[a]	444 000[e, f, k]
	2 700[h]	33[e]	
Tigris	1 658[k]	42[f]	373 000[g]
	1 840[a]	49.2[a]	378 850[e]
	1 900[g, h]	47[e]	471 606[k]
Jordan	320[d]	1.3[c]	11 500[d]
	360[e]	1.5b	16 000[g]
	c. 800[a]	1.6[a]	18 300[i]
		1.85[f]	19 850[e]
Nile	6 825[a]	83.6[a]	2 900 000[d]
			3 031 000[e]

Sources: a: Lowi (1993a: 29); b: Kolars (1992: 110); c: Beschorner (1992: 8); d: Bulloch and Darwish (1993: 40); e: Gleick (1994a: 13); f: Starr and Stoll (1988: 143); g: Grolier (1993); h: Hillel (1994: 92); i: Murakami and Musiake (1994: k: E. Anderson (1991b: 13)

Note: One billion cubic metres (bcm) equals one km^3 or 1×10^9 cubic metres

water as a conflict issue emerged within the context of an extensive ethnic and political antagonism – Arab–Israeli conflict – in which water has been more a pretext than a cause of interstate strife; and that the examples often presented as evidence of water wars, do not stand up to detailed analysis. We shall also show that, although political relations between the riparian countries have sometimes been negatively affected by disputes over water rights, these water disputes have often tended to encourage cooperation rather than hostility. Furthermore, we shall indicate that even during the time of bitterest hostilities between Arabs and Israelis, there was a tacit, albeit limited, cooperation over water. Regarding recent developments in the region, we shall argue that the new generation of policy-makers in the region, particularly the new generation of Jews who were born and bred in Israel, are well aware that conflict cannot either change the geographical givens or resolve the water crisis (Peres with Naor 1993: 128), and a vision of environmentally sustainable development in the context of regional economic cooperation is emerging as an alternative to the sterile conflicts of the past. This prospect may seem remote to sceptics but the seeds of future cooperation have been planted in many minds and it is certainly in evidence in the emerging approach to water issues.[2]

Map 4.1 The Jordan River basin

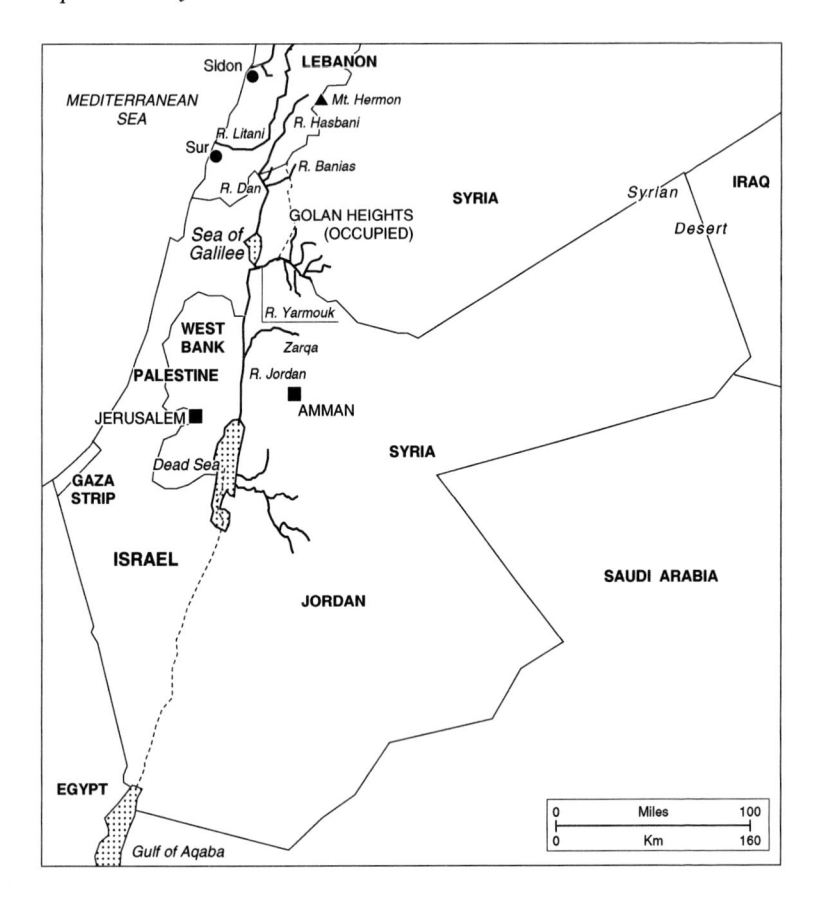

Climatic and hydropolitical features

Mountains, dry plateaus, and deserts dominate the landscape of the Middle East. Except for a mountainous belt which stretches across the northern part of the region, the region is generally characterised by aridity and contains large desert and semi-desert areas with small islands of well-watered lands.

The Jordan River Basin is a zone in the Middle East which has two kinds of very different climatic characteristics. First, the environment shared by the riparian countries of the Jordan River experiences a sharply varying seasonal rainfall – two distinct seasons predominate in the basin: a rainy period from November through March; and a dry season which extends over the next seven months. Second, there is a marked spatial disparity in the distribution of precipitation over the basin – annual precipitation ranges from less than 25 mm/ year in the southern parts of Israel to more than 1400 mm/year in the mountainous areas of Lebanon and Syria (Murakami and Musiake 1994: 118; Israel Information Service). This climatological feature confines about 60 per cent of Israel, 50–60 per cent of Syria, and more than 85 per cent of Jordan to the arid or hyper-arid desert which receives negligible amounts of rainfall, and restricts the main recharge areas of the basin to its northern belt and the hills of the West Bank. Furthermore, consecutive years of severe drought have been a well-known phenomenon in the region since biblical times.[3]

The Jordan River consists of four principal tributaries originating in four countries (see Map 4.1).[4] Three spring-fed tributaries generate the Upper Jordan: the Dan in Israel, the Hasbani in Lebanon, and the Banias in Syria. As Table 4.2 illustrates, the Dan, which is the largest tributary, has a relatively steady flow of around 245 million m^3 per year (mcm/yr), nearly 50 per cent of the discharge of the upper Jordan. The Hasbani, which on average contributes about 138 mcm each year to the river flow, is subject to sharp seasonal and annual variation – between 52 and 236 mcm/year. The Banias, which is the smallest river, originates in Syria, less than 2 kilometres from the Israeli border. Its average contribution to the Upper Jordan is 121 mcm/year and, like the Hasbani, it is subject to extreme irregularity (Naff and Matson 1984: 17–19). The Upper Jordan flows southward into the Sea of Galilee, the only natural freshwater reservoir within the basin. Not more than 10 kilometres down from the Sea of Galilee, the Jordan is reached by the Yarmouk. This is its most important tributary, and contributes 400–500 mcm of water to the basin (Naff and Matson 1984: 20; Lowi 1993a: 28; Gleick 1994a: 9). While the quality of water in the Upper Jordan River is very good, the Lower Jordan is mostly saline and by the time it enters the Dead Sea the water, if any, is too salty to use.

The Jordan basin includes Syria, Lebanon, Israel, the Kingdom of Jordan, and the Occupied Palestinian Territories. Among these countries, Israel, Jordan, and Palestinians are suffering from an accumulating water deficit and they are the most dependent riparians on the Jordan River waters. These three parties make up the core of the basin for three reasons. First, they incorporate the major part of the catchment area.[5] Second, they do not have other surface

Table 4.2 Major tributaries of the Jordan River

Tributary	Headwaters	Annual discharge (mcm)
Dan	Israel	245–260
Hasbani	Lebanon	117–138
Banias	Syria	121–125
Hullah springs and local run off	Israel	180
Yarmouk	Syria	400–500[a]

Sources: Naff and Matson (1984: 17–21); Lowi (1993a: 28); Kliot (1994: 175–80)

Note: a. Lowi (400 mcm), Naff and Matson (500 mcm)

water resources, and only limited groundwater sources, a fact that makes them particularly interested in the waters of the Jordan River basin. And, third, they are historically the primary users of the basin's waters – according to Naff and Matson (1984: 27), only about 5 per cent of the total water demand of Lebanon and Syria is satisfied from the Jordan basin, while the Upper Jordan proper supplies Israel with about one-third of its total water consumption. The Kingdom of Jordan historically satisfied its major water needs from the River Jordan and its tributary, Yarmouk. Although after the Six Day War in 1967, Jordan lost much of its share to Israel, it still uses the river to provide for about 50 per cent of its water requirements (Lowi 1993a: 20).

Syria and Lebanon, because of their claim on two headwaters of the Jordan River, are an integral part of the basin, yet the main part of their territories, and their most important agricultural areas, are fed by other river systems of far more importance. Syria is relatively well endowed with both surface and sub-surface water resources (Gischler 1979: 113). As Table 4.3 indicates, Syria's total water supply is about 36 billion cubic metres (bcm), sufficient for its immediate and foreseeable needs. The bulk of Syria's water demand is covered by the Euphrates, which crosses the country from north to east, and by the Orontes which nourishes the northwest of the country. Currently, the River Yarmouk provides for only 7 per cent of Syria's total water needs and irrigates about 6 per cent of its agricultural land (*Syria Statistical Abstracts* 1990). Lebanon, for its part, is a mountainous country and almost all of its territory receives adequate precipitation. Moreover, there are several rich internal rivers which create a significant water surplus for the time being and for Lebanon's anticipated needs (see Table 4.3).

Table 4.3 Water supply and demand in the Jordan River basin

	Supply	Demand in 1990s	Demand in 2000
Syria	36 bcm	6 bcm	8–9 bcm
Lebanon	5 bcm	900 mcm	1.5 bcm
Jordan	850–900 mcm	1 bcm (1995)	1.12 bcm
Israel	1.5–1.7 bcm	1.9–2 bcm	2.1 bcm
Gaza Strip	60 mcm	100–120 mcm	200–50 mcm
West Bank	560–670 mcm[a] 580–710 mcm	475–500 mcm (Israel) 130 mcm (Palestinians)	?

Source: Kliot (1994: 222–48); a. Naff and Matson (1984: 47)

Though Lebanon and Syria may have more water than they require for current uses and at least for the next few decades, Jordan, Israel, and the Palestinians' overall water resources is extremely limited, unevenly distributed, and highly subject to climatic fluctuations. It is in these territories that the biblical prediction of 'seven years of drought followed by seven years of plenty' originated. And this is why in this region small rivulets such as the Jordan and the Yarmouk assume importance out of all proportion to their modest discharges.

The Kingdom of Jordan, one of the driest countries in the world, with about 90 per cent of its land receiving less than 200 mm of annual rainfall, has an average renewable water supply of 850–900 mcm/year, including groundwater, the Yarmouk, and a few other small surface sources. Being deprived of the opportunity to develop its water storage capacity on the Yarmouk River, this country has been over-utilising its underground water resources for many years, and it seems that this pace will increase in the future. In 1985, Jordanian water demand was 950 mcm: that is, 110 mcm in excess of its available supply of 840 mcm in that year. Even if Jordan could increase its water supply to 1.1 bcm by the year 2000, it is estimated that its water deficit would still range between 100 to 300 mcm (Kliot 1994: 225–30).

Located on the edge of a desert belt, Israel not only suffers from scarcity of water, but also from its maldistribution. According to Beschorner (1992: 10), the north of the country provides 80 per cent of Israel's water resources, while 65 per cent of its agricultural land and the largest cities lie in the south. Despite its efficient national planning, advanced technology and exploitation of water resources taken from its neighbours by force, Israel still faces an increasing annual water deficit. While figures vary, its total annual supply of renewable freshwater resources amount to some 1.6–1.9 bcm, of which about 75 per cent is used for irrigation and the balance for urban and industrial purposes. The country's water sources include the Jordan River, natural springs, aquifers, and seasonal local runoff (Israel Information Service; Murakami and Musiake 1994: 117). Since all freshwater sources are already overexploited, including the aquifers of the West Bank, new ways are being developed to exploit marginal water resources through wastewater recycling, cloud-seeding and desalination of brackish water (Feitelson 1996: 17; Homer-Dixon 1994a). According to Israel's State Comptroller, water utilisation has exceeded the level of renewable water sources by 200–300 mcm/year and a disastrous deficit of 2 bcm has accumulated (Kliot 1994: 232).

Decades of overpumping have caused sea water to invade Israel's coastal aquifer, a key freshwater source. The degradation of the coastal aquifer greatly deepens Israeli dependence on the aquifers underlying the West Bank. To protect this important source, since 1967 the Israeli government has strictly limited water use by Arabs on the West Bank, while overdrawing the aquifer for its own uses – an inequity that has greatly angered the Arab population (Postel 1992: 76). Palestinians' quota of 130 mcm/year represents only 20 per cent of the rechargeable groundwater reserves of the West Bank, estimated as ranging from 560 to 710 mcm (Naff and Matson 1984: 47; Kliot 1994: 247). The Palestinians demand an equitable share in the water resources of the area. However, the West Bank's main water potential is already fully exploited and Israel cannot increase the amount of water available to it within its pre-1967 borders except by desalination of sea water or importing fresh water from outside the basin.

The water resources of the West Bank will play a crucial role in any plan for the future of the Occupied Territories. Being a scarce resource in the West Bank and Gaza – as it is in Israel – water is going to be the most difficult resource to divide or share. This is clear particularly with regard to the water balance in the Gaza Strip, in which renewable water resources are 60 mcm whereas demand is 100–120 mcm/year. The gap between supply and demand is met by overpumping, which has led to a drop in the water table and an increasing salinisation of wells. Over-utilisation of the underground sources has continued in Gaza for more than 30 years, as has the process of salinisation (Kliot 1994: 244). Moreover, as indicated in Table 4.3, water demand in the area will reach 200–50 mcm by the year 2000 (Kally 1986: 75), by which time the Gaza

Strip will have a total water deficit of about 140–190 mcm – which enormous amount will have to be met by importing water from outside the region and/or by desalinating seawater (Kliot 1994: 245).

The above-mentioned hydropolitical configuration implies that the water crisis in the Jordan River basin is a matter of both absolute scarcity and inequitable resource distribution among the riparians of the shared water resources. As far as the problem of absolute scarcity is concerned, it is a phenomenon as old as the history of the region (see Chapter 1). However, indigenous people were well adapted to their natural environment and, up to the end of the period of Ottoman rule, the Jordan system was sustainably utilised for local irrigation schemes. According to Naff and Matson, 'Serious problems arose when Jewish immigrants started to arrive in large numbers early in the British Mandate. . . . The water issue became more urgent in the 1930s when Jewish immigration to Palestine increased' (1984: 30, 32).

In fact, apportionment of the region's water resources only became a matter of dispute in the context of Arab–Israeli conflict which is, according to Homer-Dixon (1994a: 2), an ethnocentric or 'group-identity' conflict. In this type of conflict, 'each group emphasizes its own collective identity – its "we-ness" – while denigrating, discriminating against, and attacking members of the other group' (Homer-Dixon 1991a: 105). The group that has sufficient power and sufficiently authoritarian institutions, will often try to control the other group's population size through deportation, exclusion, or extermination, a process which leads to a 'population race' as each group tries to increase its own population faster than the other (Homer-Dixon 1994a: 2). As we shall see later, this has been exactly the case in the Arab–Israeli conflict, and it has had a severe effect on the water balance in the region.

Since the struggle for the possession of land and water has been the two-pronged basis of the Arab–Israeli conflict, and since the conflict began in this basin as a result of the rise of Zionism with its aim of establishing a Jewish state in Palestine, let us start our analysis of 'water politics in the Jordan River basin' by a brief review of the role of Zionism in the growth of the water crisis in this region.

Evolution of the water crisis in the Jordan basin – the role of Zionism

Historical background

Understanding the dynamics of the water crisis in the Jordan River basin and its security, economic and foreign policy implications, necessitates some historical background. From time immemorial, the lack of the natural resource of water has been a cardinal environmental issue that has shaped the overall structure of civilisation in the region and deeply influenced the indigenous societies of the basin (see Chapter 1 and Ngan 1991). However, as far as foreign policy is concerned, the issue of a water crisis is a new phenomenon in this region. The reason is that from the time of the Assyrian Empire – 9th century BCE – up to 1918 the land of Palestine and its periphery were subject to rule by a succession of large empires. Assyrians, Babylonians, Persians, Greeks, and Romans in turn conquered Palestine, which fell to the Muslim Arabs by 640 CE. Afterwards, the area was the focus of the Crusades and eventually was conquered in 1516 by the Ottoman Turks, who governed the territory until the First World War. By the late nineteenth century, Zionism had arisen with the aim of establishing a Jewish homeland in Palestine, and during the First World War the British, who captured the area, appeared to support this goal.

Up to this time, there was no international water conflict in the region. The postwar settlement, however, partitioned the formerly unified basin of the Jordan river between British and French Mandates. This process was carried further when the Mandates became independent. Just before the British conquered Palestine in 1917, its total population was about 600 000 (Eaton and Eaton 1994: 94; Hillel 1994: 84). Despite the aridity of the climate, the relatively few indigenous residents of the basin were accustomed to their natural environment and over many centuries had adjusted their lifestyle accordingly. By contrast, when the Zionist movement started settling the Jews in Palestine, water supply was a vital factor in their aspirations. From the outset, access to water was a prerequisite for the settlement of the newcomers, who sought to adjust the natural environment to their previous lifestyle, and unrestricted access to water resources was perceived as a non-negotiable prerequisite for the survival of a Jewish national home (Lowi 1993a: 40); it was of central importance in the Zionists' quest for land and for delimiting the desired territories well beyond the 'biblical' promises and claims[6] (Perera 1981: 54; Zarour and Isaac 1993: 41).

The Jordan River Basin 95

Soon after the First World War in 1918, while the borders of the British Mandate of Palestine were still under dispute between the British and French governments, the Zionist delegation to the Peace Conference in Versailles was actively lobbying in favour of including the entire watershed of the Jordan and Yarmouk rivers within the borders of the national home for the Jews, to embrace all the Palestine lands, southern parts of Lebanon and Syria, and the Jordan valley proper.[7] The Zionist leaders also demanded the incorporation of the lower section of the Litani River – a river in the south of Lebanon – the waters of which were to be diverted into the Hasbani River through a short tunnel to the Jordan River (Hurewitz 1956: 45–8; Hewedy 1989: 23–5; Eaton and Eaton 1994: 95). When the Zionist leaders failed to secure the Litani and the Jordan's headwaters, they tried to establish settlements in Syria and Lebanon, but the opposition of the French thwarted these attempts as well (Nijim 1990: 318).

The French position, however, did not prevent comprehensive hydrological investigations and assessments in the region. The surveys were carried out to estimate the available water resources in the basin and to suggest the best method for optimal utilisation of the Jordan water system. Among scores of studies which were conducted in this field (see Naff and Matson 1984: 31 and Kliot 1994: 189–98), one stated that 'the Zionists welcomed Lowdermilk's plan and considered it as their water constitution' (Mustafa 1994: 125). Having a 'Tennessee Valley Water Authority' approach to the Jordan River basin, Lowdermilk outlined a basin-wide comprehensive plan in 1944. He proposed using the waters of the Jordan River and all its tributaries (Yarmouk, Banias, Hasbani, Dan, and Zarqa – a river in East Jordan), to irrigate the Jordan Valley, much of Northern Galilee, and Northern Palestine. He also suggested diverting the Litani River to form an artificial lake whose waters would be pumped southward to irrigate the Negev Desert. An early version of a canal between the Mediterranean and the Dead Sea was also included in his proposals (Kliot 1994: 190).

It was quite clear that cooperation between Arabs and Jews was a prerequisite for the creation of the 'Jordan Valley Authority'. However, due to severe ethnocentric hostilities, no cooperative solution was found and the United Nations' partition plan of 1947 aggravated the difficulties. The Arab–Israeli war of 1948 further complicated efforts for a regional water solution (Cooley, 1984: 9). With Israeli sovereignty, came the power to control water resources, hence the Zionist leaders abandoned the idea of a region-wide development of water resources and shifted to planning at national level. The new state of Israel thus became determined to tap, develop, and exploit unilaterally any available water resource. Before discussing the ramifications of the water policy of Israel as an independent state in the basin, let us review the process of Jewish settlements in Palestine and its impact on the emergence of the water crisis in the region.[8]

Early Zionism

The dominant idea of early Zionism, as a political ideology, was to establish a Jewish 'national homeland' in Palestine and to encourage Jews to return to their 'home'. Returning to one's homeland connotes returning to one's roots, but Jews had to find their roots in an exotic and adverse environment. According to de-Shalit (1995: 74), the Jews, who mostly came from Europe or other temperate zones, were disturbed by the strange and hostile environment. The strength of the light, the summer heat, the sandstorms, the mosquitoes, the stony mountains, the marshy land and the swamps were all worrisome aspects of the alien physical environment to newcomers. Psychologically, they were overawed by coming to a land with very few facilities and an unknown future, having no friends and suffering from unemployment. Besides, they were not welcomed by the indigenous inhabitants. These obstacles seemed to cast doubt on the whole prospect of Zionism. The phenomenon of a massive immigrant population confronting an environment with which it was totally unfamiliar and which was very different from their expectations is a feature of early Jewish settlements in Palestine. The response of the Zionist leaders was twofold: first to romanticise the environment, and second, to conquer it, to make it flourish, and to make it more inhabitable with familiar colours and shapes (Ibid.). Let us look at the romanticisation process first.

The development of settlements was the most important determinant of Zionism's nation-building effort during its formative years. However, since life in Palestine was quite harsh for those who were not accustomed to the climatic conditions in the region, for Zionism to fulfil its ambition, politicisation of the environment was an urgent need. The Zionist leaders resorted to ideological arguments as an instrument to overcome the obstacles; immigration to this chosen land was deliberately given a romantic interpretation. As de-Shalit (1995: 71–3) points out, by creating myths, symbols and rituals which would support and sustain socialisation and integration within such a harsh and exotic environment, the Zionist leaders found a way to romanticise events for pioneers – the early Jews who came to Palestine. They glorified the renewal of the relationship between the Jew and the soil as a 'purifying experience', and as a liberating source of 'self-emancipation'. The romantic attitude to nature thus became almost a religion. The early immigrants, therefore:

> regarded themselves as *returning*, not only in the geographical sense – to the land of Israel – but also in the temporal sense, to the pre-exilic period, to their national and cultural roots in the age of the Bible; and they thus became farmers, cultivating the land with an ox and a plough; they lived in rural areas, away from the cities which they saw as *degenerate* and *immoral*. (de-Shalit 1995: 73)

Since settlements were an important component in the Zionist nation-building effort, for two reasons the romanticisation of the environment was a political imperative. First, it could answer a psychological need, arising from a feeling of anxiety about a completely strange, foreign, and hostile environment. Second, it was cost-effective in establishing the Jewish presence in relatively water-rich areas, many of them remote from Jewish centres and inhabited by local Arabs (Feitelson 1996: 18). Accordingly, agriculture was considered as the only way in which the immigrant Jews could overcome their anxieties and develop such strong ties to their land that they would never again be thrown out of it.

By romanticising the attitude to nature, the Zionist leaders succeeded in shifting the focus of immigrant Jews from Jewish centres such as Jerusalem and Safad to the rural areas. They concentrated on the expansion of Jewish agricultural settlements in areas with access to water (Eaton and Eaton 1994: 96). The organisations that were in charge of facilitating the return of Jews to their homeland purchased the cheapest land to be found, which was the marshy country in the valleys, and directed the immigrants to these areas.[9]

The second response of the Israeli government was to conquer the environment. The settlers were encouraged to transform this environment into one that was more familiar. They drained the soil so that it could be used for agriculture, and they planted trees to make the land of Israel look as it appeared, according to the Zionist interpretation of the biblical myths, in ancient days (de-Shalit 1995: 78).[10]

These policies successfully facilitated the expansion of the settlements in areas with access to water. After the League of Nations approved the British Mandate of Palestine in 1922, Jews immigrated in large numbers and settled mainly in such areas. As a result, when the United Nations adopted Resolution 181, partitioning Palestine in 1947 between an Arab and a Jewish state, the upper part of the Jordan River basin was already populated by a considerable number of Jews who were living in cooperative farms. These settlements were incorporated within the boundaries of the Jewish state, including Lake Hullah, the Sea of Galilee, the Jezreel Valley, and most of the fertile coastal strip (Eaton and Eaton 1994: 96). Inevitably, given the climatic conditions of the region, the expansion of agricultural activity entailed over-exploitation of the limited freshwater supplies of the basin. However, the indigenous Arabs, who were anxious about the increasing presence of the exogenous Jewish population, opposed this trend as a zero-sum game in which they were the losers. From this time onwards, water became the central issue in the Jordan River basin – a unification of hydropolitics and geopolitics. As we shall see in the next section, after the establishment of the Jewish state in 1948, and the emergence of a new phase in the Zionist approach to the environment, this unification was *institutionalised* in the context of the Arab–Israeli conflict.

After independence

Once the Jews had established their own state in their promised land, the gates of the country were thrown open and waves of Jewish immigrants swept into Israel. As Table 4.4 indicates, at the outbreak of the First World War, the Jewish population in Palestine was about 80 000, as compared to 5000 in the early 1500s. When the British Mandate over Palestine ended (14 May 1948), the Jewish population was some 600 000. However, the main waves of unrestricted immigration reached Israel after its establishment as an independent state. By the end of 1951, a total of 687 000 had arrived, thus more than doubling the Jewish population in less than four years. When Israel celebrated its tenth anniversary, the Jewish population numbered over two millions – a 400 per cent increase.

Israel's gains weighed especially heavily on the Kingdom of Jordan, which had to accommodate an influx of about 450 000 Palestine Arab refugees out of the 750 000–900 000 Palestinians who fled or were expelled from the areas of Palestine that fell under the

Table 4.4 Jewish population growth in Palestine/Israel

Early 1500s	5,000
1914–18	80 000–85 000
1948	600 000–650 000
1951	1 300 000+
1952	1 600 000+
1958	2 000 000+
1996	4 000 000+

Source: Israel Information Service; Hillel (1994: 84, 303)

sovereignty of the State of Israel. The population of the Jordanian Kingdom at the time of its separation from Palestine was less than 200 000 – mainly consisting of nomadic Bedouin tribes, settled *fellahin* (cultivators), and a few town dwellers (Hillel 1994: 87). However, by 1948 its inhabitants had increased to some 430 000, and after the independence of Israel the population of the kingdom trebled in less than two years (Lowi 1993a: 47; Naff and Matson 1984: 34). Since most of the Arab refugees were peasants, the only way to rehabilitate them was to increase the area of arable land. Because of the hydrological situation in the country, however, the possibilities of doing so were severely limited and the only alternative was the diversion of the contested waters of the Jordan River.

For Israel, the urgent necessity of absorbing so many new Jewish immigrants with different ethnic, cultural, social, and linguistic backgrounds, and the immediate needs of the independent new state for economic growth necessitated, along with the expansion of agriculture, rapid urbanisation and industrialisation. The ethos of development was initiated by the government to socialise the new immigrants and to gather all sectors of the population around one great idea – 'the transformation of the Jewish soul on the Jewish soil'. Development was also politically significant because it could symbolise a *collective* success, sustaining the sense of *togetherness* within a diverse society and proving that a new type of Jew was emerging in the land of Israel (de-Shalit 1995: 75–7).

The ethos of development materialised in three processes – afforestation, desiccation of wetlands, and industrialisation. As noted above, at the early stages of the Jewish settlements, the idea of afforestation was based on the romantic ideal of making the land of Israel look as it was assumed to be in the biblical times. It was also designed to eliminate the sense of alienation that immigrants felt towards the environment, by transforming it into one that was more familiar to them. But after the creation of the state of Israel, afforestation acquired additional functions. In the context of development, it was an easy, cheap and effective response to the problem of massive unemployment caused by the great influx of immigrants. In political terms, it was essential to appropriate the contested lands. And, from a security point of view, it was a compelling necessity to establish a permanent Jewish presence on these territories. Accordingly, in 1949, Ben-Gurion, Israel's first prime minister, ordered the plantation of a billion trees in a decade arguing that 'if we bring in dozens of thousands of Jews and change the landscape, bring life to the desert landscape... they will never be able to take us out of here' (Weitz, in de-Shalit 1995: 85). As a result of this policy, by the end of the country's first decade some 50 000 acres of land had been afforested, and trees had been planted along almost 500 miles of highway (Israel Information Service).

Desiccation of wetlands was the second major process of the ethos of development. Although this process began as early as the first immigrants came to settle in Palestine, it was in 1951 that the Zionist technological effort to overcome the alien environment reached its climax as the Israeli government decided to drain the Hullah swamp, north of the Sea of Galilee (see Map 4.2). According to de-Shalit (1995: 79), there were three justifications for this decision: to defeat malaria; to liberate the soil to make room for more settlements in the area, and to reclaim both water and soil for agricultural purposes. However, de-Shalit claims that malaria was uprooted before the swamp was drained off; the soil appeared to be agriculturally worthless; and the hydrological claim that the Sea of Galilee, the main source of water supply in Israel, would gain from the project, proved to be fallacious. Draining of the Hullah wetlands and increased agricultural activity in the area caused sediments, nutrients, and pesticides to flow directly into the Sea of Galilee and to seriously contaminate this vital freshwater reservoir (Wollman 1996: 4).

Although this project led to an ecological disaster, de-Shalit claims that:

> for many years the disappearance of the swamp could not be criticised, because the government and the ruling party represented it in their propaganda as a successful national effort, proving that the Jews in Israel could overcome the alien environment which they found when they returned to their land. Thus, 165,000

Map 4.2   The Hullah wetlands

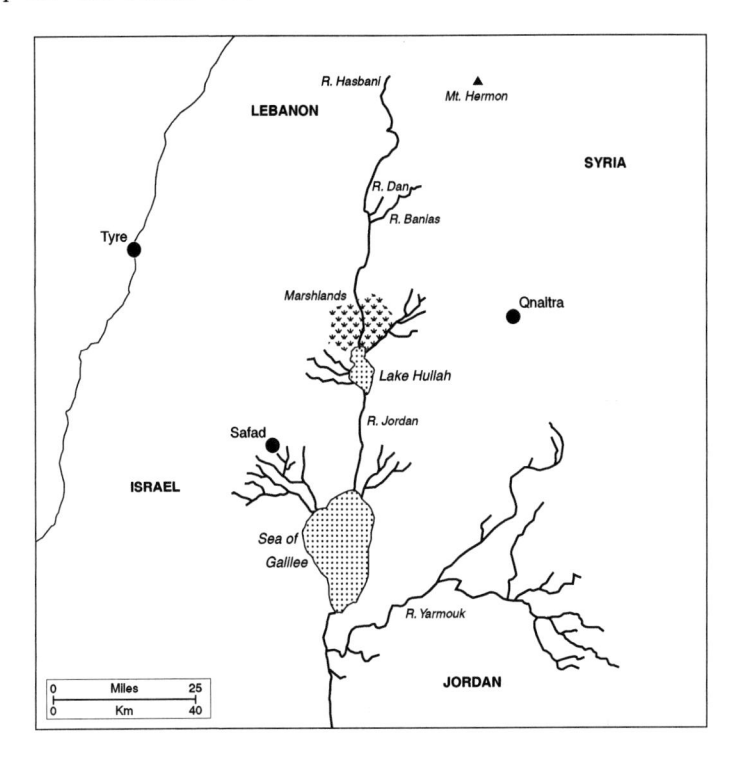

acres of natural environment were manipulated for political reasons; but although it was soon apparent that the project was a mistake, it was only much later, in the 1980s, that the JNF [Jewish National Fund] issued a new programme for the area, including re-flooding at least a third of it, at the cost of $30 million. (de-Shalit 1995: 79)

Despite its negative environmental ramifications, the project succeeded in populating the area. Indeed, it was an important phase of Israel's population and settlement policies. In conjunction with the third process of the ethos of development, rapid industrialisation (and accompanying urbanisation), and the introduction of private farming, this project severely modified the hydrological balance of the basin. These policies, particularly the introduction of private farming which changed the water-intensive agriculture from a romantic occupation to an economic enterprise, siphoned off the limited water resources of the country and created an unprecedented water crisis in the basin. According to Kliot (1994: 240), Israel's water demand in 1947 was only 230 mcm; in 1975 it amounted to 1.73 bcm and in 1989 it was 1.91 bcm – of which 1.3 bcm was used for agriculture, 495 mcm for the domestic sector, and 107 mcm for industry.

Avner de-Shalit, who is from a new generation of Israeli scholars, sums up the role of Zionism in the creation of the water crisis in the Jordan River basin, by underlining its technological and exploitative approach to the environment:

> And thus Zionism became synonymous with the conquest of the desolate wilderness. The prevailing idea was that the wilderness was bad, and the conquest of it was good, in fact the very essence of the good. The best environment, in the ethical sense, was the one in which nature had become 'human', in which the desert was abolished and there were no untamed values, but rather human values. (de-Shalit 1995: 80)

Besides their instrumental approach, the Zionists' perception of security was another factor which adversely affected the environment. Being overwhelmed by ethnocentric hostilities and territorial conflicts, the particular distribution of the Jewish population was of especial concern to the Zionist leaders and their perception of the security needs of the newly founded state. As Lowi (1993a: 33) points out, a settlement strategy was devised, primarily based on national security considerations, to implement a spatial pattern of population dispersal. It was a deliberate policy to distribute Jewish settlements throughout the country and especially in the border areas. This strategy was clearly manifested in the words of David Ben-Gurion, Israel's first prime minister, who said that 'we must go to the borders or the borders will come to us' (quoted in Mandelbaum 1988: 272, n. 26). In the next section, we will discuss the effects of this strategy on Israel's water policy and the formation of a strategic perception of the water resources in the region.

Israel's water policy and Arab reaction

Conception of water as a strategic issue

As explained above, to accommodate unrestricted Jewish immigration to the land of Israel from the early days of independence, development towns and agricultural settlements were built. To disperse the population throughout the country and to promote a closely interlocked rural and urban economy, industry and services were brought into previously unpopulated areas (Israel Information Service). Over 250 of the new settlements which were established within five years of Israel's independence (Feitelson 1996: 18) were considered as essential for security purposes – used as a first step in the consolidation of territory, providing frontier resistance, and thus valuable time, in case of attack.[11] Expansion of the settlements, mostly for security reasons and without adequate regard to existing water resources, created a new situation of massive water demand in a short time. This demand for water was not concealed – on the contrary, the Zionists' ideological aspiration to 'make the desert bloom' was explicitly linked to accessibility of adequate water resources. According to their own metaphor, water was the 'lifeblood' and 'a prerequisite for a new society' and a 'nation rooted in its land'. In his speech to the Knesset in 1952, the prime minister, Moshe Sharett, declared that:

> Water for us is life itself. It is food for the people – and not food alone. Without large scale irrigation projects we shall not achieve high productivity, balancing the economy or economic independence. For without irrigation we shall not produce a wealthy agriculture ... we shall not be people rooted in the land, secure in its existence and stable in its character. (Feitelson and Haddad 1994: 73)

Unsurprisingly, the local water resources were insufficient to meet the needs envisaged by Israeli plans, and the only alternative, the Jordan River system, figured significantly in Israel's development plans (Lowi 1993a: 49). However, the Jordan watershed was now shared by four sovereign states and, because of the prevailing ethnocentric hostilities and territorial conflicts, development of its water resources was perceived by upstream Arabs to be a zero-sum game; Arab waters were sources upon which the development projects of their bitter enemy, Israel, depended.

Conflict, therefore, between Israel and the Arabs seemed inevitable. On the one hand, Israel was determined to change the hydrological map of the basin so that it could ensure enough supplies for all the Jews who wanted to settle in Israel – the ingathering that was the main reason for the establishment of the Jewish state. On the other hand, the Arabs, having realised the sensitivity and importance of water in Israeli development plans, decided to use their geopolitical advantage of being in an upstream position as a leverage to exert control on Israel. Having their hand on the tap, the Arabs assumed that if they could limit Israel's water supplies, they could restrict the ability of Israel to absorb more immigrant Jews – a point of central concern for the Arabs in their ethnic and territorial conflict with Israel.

So water had now gained a security dimension. Having acquired an ideological, demographic, and economic significance, the water question had now become a strategic issue – a matter of national security and foreign policy. The failure of several proposals concerning the regional water-sharing arrangements in this period confirms the fact that water was involved in foreign policy considerations of both Arabs and Israelis (Dolatyar 1995). The strategic perception of water resources in a conflict-ridden atmosphere made it impossible for Arabs and Israelis to negotiate a voluntary and mutually acceptable agreement. Therefore, Israel proceeded unilaterally with its water development plans, and this provoked the Arabs to resort to countervailing action.

Institutionalisation of water policies

Since the existing water supply systems were no longer able to satisfy the accelerating demand for water, particularly the needs of water-intensive agriculture, inter-basin water transfers for Israel were inevitable. After the establishment of the Jewish state, Israel asserted state ownership of all natural resources under its control. This legislation abolished *de facto* ownership and control by landowners or villagers over local wells (Eaton and Eaton 1994: 97; Feitelson 1996: 18) in order to provide the legal framework for the operation of a national water system and thereby overcome the extreme imbalances in water availability across the country. Its central artery, the National Water Carrier, completed in 1964, brought water from the north and central regions, through a network of giant pipes, aqueducts, open canals, reservoirs, tunnels, dams and pumping stations, to the country's urban centres and to the agricultural settlements as far as the Negev Desert (Israel Information Service). To the Arabs, Israel's National Water Carrier became a symbol of aggressive expansionism, to which they responded with their own diversion plans (Cooley 1984: 14–18). Israel's perception of water as a strategic issue made its leaders consider the Arab project a grave danger to their national security and decided to use all their means, including military, to impede it. This Israeli overreaction to the water issue led Arabs to presume that they had found the Achilles' heel of their enemy. On that assumption, the first military action of the Palestinian National Liberation Movement (Al-Fatah) targeted the Israeli National Water Carrier (Zarour and Isaac 1993: 41), and the Arab League decided to help Syria, Lebanon, and Jordan to divert the headwaters of the Jordan River. When Israel used its military power to hit the Syrian construction sites, a series of continued skirmishes occurred which eventually led to war in June 1967. Emphasising the security dimension of the water issue, some Israeli politicians believed that after its battle for existence in 1948, it was time for Israel to fight for a secure access to water resources. Ariel Sharon, later Israel's defence minister explained that 'people generally regard 5 June, 1967 as the day the Six-Day War began, that is the official date. But in reality, it started two and a half years earlier, on the day Israel decided to act against the diversion of the Jordan' (Bulloch 1993: 12). However, as we shall see later, in a context in which the hostile parties were in conflict over every issue, it is too simplistic to describe the water issue as *the cause* of war. Moreover, as Wolf (1997: 6) points out, there was a time lag of twelve months after the initial Israeli military attacks on Syrian construction sites before the outbreak of the Six Day War in 1967.

Whatever the cause of the Six Day War between the Arabs and Jews, victory enabled Israel to fulfil much of its strategy aimed at controlling the headwaters of the Jordan River in Southern Lebanon and the Syrian Golan Heights. At this point, Israel achieved an unparalleled integration of its water resources into one national system. This institutional structure gave absolute authority to a central state institution and reinforced the domination of the agricultural sector. For reasons of ideological commitment to agriculture and security-demographic considerations (discussed above), Israel launched an ambitious plan of agricultural development through which water for irrigation was heavily subsidised[12] and major national efforts were geared toward increasing agricultural production, as can be seen in the increase in irrigated area set out in Table 4.5.

Table 4.5 Agricultural development in Israel

	Irrigation area (000s ha)	Water use (mcm/year)
1949	30	180
1955	89	760
1960	130.5	1060
1965	150.8	1153
1970	172.4	1233
1975	182.5	1297
1977	186.5	1258
1988	217	c. 1300
1990	235.2	c. 1200

Sources: Lowi (1993a: 31); (Israel Information Service)

While the National Water Carrier alleviated Israel's water crisis during the 1960s, a population boom brought on by increased immigration and improved health services added more demands on Israel's water supply (deShazo and Sutherlin 1994: 1–2). For Israel, the central problem following the 1967 war was to preserve its diminishing domestic water supplies and to make efficient use of water resources captured from its neighbours, of which water in the Arab farmers' wells in the West Bank became a key element. From the beginning of the Israeli occupation, military authorities prohibited West Bank Arabs from drilling new wells without special permission which, according to Cooley, was almost impossible to obtain. The occupying authorities likewise blocked or sealed many existing wells (Cooley 1984; Naff and Matson 1984; Zarour and Isaac 1993; deShazo and Sutherlin 1994: 1–2) and determined Arabs' access to water by a very restrictive consumption quota. This policy severely curtailed the growth of both agriculture and industry in the occupied territories. In a report to the Center for Strategic and International Studies at Washington DC, Joyce Starr and Daniel Stoll held that 'the main water potential of the West Bank is exploited in a ratio of 4.5 per cent to the West Bank and 95.5 per cent to Israel' (Starr and Stoll 1987: 7; see also Bulloch and Darwish 1993: 35). Over the years, the Israeli policy-makers attempted to institutionalise this inequitable mode of resource distribution in order to satisfy their needs. This attitude was clearly manifested during the Israeli–Palestinian negotiations in the context of the so-called peace process: the Israeli foreign minister Shimon Peres told Yasser Arafat in July 1995 that 'we are now negotiating over the levels of water each side needs, not about water rights' (Butler 1995: 34).

However, as Israel sought to exploit all accessible water resources, its agriculture became ever more dependent upon irrigation, and demands for water steadily increased. Irrigation in Israel's crop-producing region grew from 15 per cent in 1950 to about 64.2 per cent by the late 1980s (Beaumont 1989: 384; World Bank 1994: 70). It is true that during this period the average water use per unit area fell by one-third with the introduction of more efficient water application techniques, such as drip irrigation. Nevertheless, water demand rose as more immigrants arrived, more land was cultivated, and the population of the country grew. The situation in other riparian nations was the same as in Israel; ever-growing population, expansion of agriculture, and over-utilisation of the drainage basin. As a result, the Jordan River has been reduced to the 'carrier of a saline trickle to the Dead Sea' (Pearce and Hudson 1991). By the early 1980s it was clear that the Negev Desert reclamation could not be achieved without additional water resources. Given water scarcity, the high costs of desalination and other unconventional methods of supplying water, and the pollution of surface and ground waters, Israel needed other natural supplies of water, and Lebanon was the only neighbouring country with surplus water resources. When in 1982 Israel invaded Lebanon under the pretext of security, some commentators such as Hewedy claimed that:

> the broader incentive for the invasion of Lebanon in 1982 was to secure the waters of the Litani River. The avowed objective to protect the northern borders from Palestinian attacks is not convincing to most military observers. Rather, the thrust appears directed at finally seizing control of the Litani River. (Hewedy 1989: 23)

Moreover, many writers view Israel's retention of Southern Lebanon as an extension of its persistent efforts to secure the Litani waters – a view implied in Moshe Sharett's entry in his diary: 'The Israeli army will enter Lebanon, will occupy the necessary territory and will create a Christian regime that will ally itself with Israel. The territory from the Litani southwards will be totally annexed to Israel and everything will be all right' (Dillman 1989: 56). However, Alam (1998: 6) explains how interviews with Israeli generals indicated that water was not a major factor: construction of desalination plants safely behind Israeli's borders would have been cheaper than the invasion.

Sharett's dream did not come true. Because of the unexpected resistance it faced in Lebanon, Israel was forced to withdraw its forces. But its need for the Litani river made Israel establish a security zone to secure access to the water. The amount of water that Israel diverts from the Litani is a matter of controversy, but Israel's planners have concluded that the appropriation of the Litani waters even to the extent of adding some 500 mcm to Israel's water supplies can be conceived as only a short-term solution, given the present structure of its water policy and economy, and given the average increase of Israel's water consumption by one per cent per year (Hewedy 1989: 25). Indeed, they have admitted that present supplies can scarcely meet the country's current needs, let alone the expected level of consumption beyond 2000.

The new paradigm

Three major events have combined to challenge Israel's efforts to institutionalise its water policies in accordance with a strategic perception of the issue, shifting the emphasis away from competition and confrontation, towards potential cooperation. The first event was the long drought in the 1980s, as years of below-average rainfall generated an atmosphere of emergency among policy-makers. Water tables dropped; shallow wells dried, and the Sea of Galilee – which had been supplying almost one-third of Israel's demands – fell to its lowest level for 60 years. The State Comptroller, Miriam Ben Porat, issued a special report in January 1991, confirming that 'in practical terms, Israel has no water reserves in its reservoirs.' Ben Porat blamed the agriculture ministry for allocating too much water to farmers while ignoring warnings of shortages, and she claimed that 'Today, there is a real danger that it will be impossible to provide water in enough quantity and quality even in the short term' (quoted in Starr 1991: 25). This situation caused a dramatic tightening in Israel's water management practices, including rationing, cut-backs to agriculture, restructuring of water pricing and allocations – that is, economic solutions.

This economic approach to the water issue reflected the fact, reported by Feitelson (1996: 17), that 'an increasing number of Israelis, and especially in the professional elites, accept today the view of water as a commodity, rather than a strategic resource.' This change in attitude toward water, Feitelson argues, can facilitate water negotiations, as 'Israelis come to realise that relinquishing some water is not a life-or-death issue, but rather a financial one.' This is

Table 4.6 The position of agriculture in the economic structure of Israel

Branch	GNP	Labour force	Exports[a]	Investments
Industry	24	22	62.4	24
Agriculture	3	4	2.4	2
Construction	7	7	–	30
Transportation and communications	9	6	10.6	7
Commercial, financial and personal services	28	32	24.6	7
Public services	29	29	–	13

[a]Includes goods and services

Source: Bank of Israel, *Annual Report, 1993* (Israel Information Service)

partly because Israel has become industrialised; from the early 1980s, the agricultural sector has become so marginalised that by 1993 its share in the gross national product (GNP) had declined to 3 per cent, while the proportion of the Israeli labour force engaged in this sector declined from 17.3 per cent in 1960 and 8.8 per cent in 1970 to 4 per cent in the early 1990s (see Table 4.6).

While Israeli agriculture has slowly been moving from the wet north into the dry southern areas, due to increasing urbanisation in the north, 'the children and grandchildren of the Labor Zionists who plowed the land and made the desert bloom, ask why should we bother growing our own food?' (Starr 1995: 65). This economic approach was not without its critics, however. Indeed, as Joyce Starr reveals, 'the internal war between those who believe in relative food self-sufficiency and those who hope to eliminate agriculture altogether is one of Israel's few well-kept secrets' (Ibid.).

Since a high portion of the water is used for export cultivation, the State Comptroller declared that the water subsidy is in effect passed on to foreign consumers, describing it as 'exporting water at a loss'. Criticising heavily subsidised agricultural exports, Israeli officials are quoted as saying, 'Don't send vegetables to Europe, just send water!' (Starr 1995: 80). This reflects the much more fundamental shift toward market-orientated mechanisms by new political and social forces in Israel. It is of special importance in the case of water policies as it marks a shift from viewing water as a strategic resource, toward a view of water as a factor of production or commodity which has an economic value. This is why for the first time marginal prices for irrigation were raised to a level which could affect farmers' behaviour, though they are still below production cost (Feitelson 1996: 20).

The idea of water as an economic factor that should be priced according to economic considerations, has been promoted by Israeli and Arab economists on the ground that the value of water under contention between Israel and her neighbours is small when compared to the cost of conflicts. According to Feitelson (1996: 20), the sums of money involved are less than half a per cent of Israeli GNP and thus are not of much consequence from a macroeconomic perspective. The view of an increasing number of professionals within both Israel and the Arab world is that the water question is primarily one of pricing and financing, and that water trading can be used as the allocating mechanism of water between Israel and the Palestinians, leading to an efficient allocation of water to both parties (for example, see Ben-Shahar *et al.* 1989; Zarour and Isaac 1994; Zeitouni *et al.* 1994; Feitelson 1996). This would arguably reduce much of the current tension over water, as the amounts of water demanded by both sides would decline.

The second major event challenging Israel's strategic view of water resources, is the rising environmental awareness in Israel, along with its promotion at an international level. According to de-Shalit (1995: 80) and Feitelson (1996: 19), until the late 1980s, the ethos of development outweighed environmental values and considerations. However, among both the general public and governmental bodies in Israel a new concern for the environment has emerged, a process which followed the establishment of a ministry for environmental issues in 1988. As de-Shalit observes, it seems quite likely that today's generation will make significant changes in the Zionist attitude to the desert, water, and the natural environment in general. He argues that:

> These people were born to this landscape; they are familiar with it and need not fear it. For this generation the desert may very well be part of daily life. Zionism at the threshold of the twenty-first century is beginning to realise and acknowledge that the environment is a given fact, and it is here to live with, neither as an enemy nor as a means of emancipation, but as it is. (1995: 81)

There are two reasons for this environmental optimism. First, the decline of the ideological, strategic, and economic significance of agriculture has forced this sector to reduce considerably its water consumption – by 400 mcm since 1986, according to Allan (1996: 113). Indeed, all the factors which led to agriculture being seen as a priority sector have changed in significance: agriculture is not a romantic activity any more; food security is no longer associated with the export-orientated agriculture; the Israeli borders of pre-1967 seem today secure and accepted by its neighbours. Moreover, as Table 4.6 indicates, the marginal share of agriculture in the labour force, and the structure of this force, which is composed mainly of Palestinian or temporary workers, have virtually eliminated the ideological priority of farming (Feitelson 1996: 20).

Second, as Israel's geographical demarcation grows clearer and becomes unequivocal as a result of the peace process between Israel and its neighbours, the environment becomes less abstract and its political interpretation approximates to the scientific description of an ecological system. Growing attention to environmental issues in the context of ecological understanding gives rise to an environmental perspective from which both Israelis and Arabs face a common threat, as pollution or over-exploitation may damage the aquifers both parties share, to the detriment of future generations of both sides. To address this common concern the two parties clearly need to cooperate. From this perspective joint management of water resources is warranted: when viewed from an environmental perspective, water negotiations are no longer a zero-sum game (Feitelson 1996: 17).

The third major event was world-wide in nature, but it clearly crystallised in the Middle East. The end of the Cold War and the collapse of the Soviet Union, together with the Kuwait Crisis of 1990-1, caused a shift of political alliances in the Middle East and made most Arab countries reassess their attitude toward Israel. These developments finally enabled the first public face-to-face peace talks between Arabs and Israelis which took place in Madrid on 30 October 1991. So far, in both bilateral and multilateral negotiations, water and environmental issues have been among the most important key points. For example, the peace treaty between Israel and Jordan, signed on 26 October 1994, includes five annexes, two of which address water and environmental issues (Israel Foreign Ministry 1994). Since the Madrid negotiations, and since the launch of multilateral talks in Moscow in January 1992, there have been numerous seminars, conferences, and papers concerning the scarcity of water resources in the Middle East and the impact of that scarcity on the region's politics. Discounting the political rhetoric from all sides, many writers, analysts, and officials have confirmed the fact that given the natural circumstances, 'multilateral co-operation' is the region's only choice to avoid the coming water crisis. For instance, expressing his anxiety about the future, Shimon Peres declared that 'the water shortage proves the objective necessity of establishing a regional system.' He rightly confirmed that 'like all wars in the political and strategic reality of our times, wars fought over water do not solve anything. Gunfire will not drill wells to irrigate the thirsty land, and after the dust of war has settled, the original problems remain. No war can change geographical givens' (Peres with Naor 1993: 127).

Conclusion

The message of this chapter is that it is the attitudes and behaviour of humankind toward water that will in the end determine how well water problems are dealt with and whether they will be resolved peacefully or by conflict. The Jordan River basin is an ideal model for studying the most acute aspects of the world's freshwater issues in their physical, social, and political presentations. This is because, since ancient times, water has been a central factor in this basin and continues to be central today. The evolution of the water crisis in the Jordan catchment outlined in this chapter indicates that the massive influx of immigrants to the region with a hostile attitude to the environment caused severe social and environmental disturbances. However, the Arab–Israeli conflict did not revolve around rights to the use of any given resource, not even water resources. It was rather concentrated on the far more profound issue of the very right of Israel to exist as a sovereign state and the opposed rights of the Palestinians to nationhood, though water resources and their exploitation have always been a subject of dispute in that context.

Even the Six Day War was not caused by water scarcity; the water issue was simply one (important) element in a wider dispute. As Libiszewski puts it: the 'hydraulic imperative' analyses of the Israeli invasions of 1967 and 1982 are 'too simplistic' – though he is willing to admit that water-related considerations may be the cause of the retention by Israeli of certain occupied Arab territories. Libiszewski argues that the Arab–Israeli conflict 'was triggered by the endeavour of Zionism to build a Jewish state on the land of their historical ancestors, and by the rejection of it by the indigenous Palestinian population and the neighbouring Arab states', and disputes over water 'were rather an outflow of political and territorial conflict than part of its origin'. Regarding the underlying cause of conflict in the Arab–Israeli and Israeli–Palestinian context, Libiszewski rightly maintains that water is *not* the cause. Rather, he argues, 'the conflict is over national identity and existence, territory, as well as over power and national security'. Even as a channel for expressing conflict, he denies that water has 'any noteworthy connection to the channels within which the Arab-Israeli conflict is being and has been fought out'. On the overall role of water as a catalyst of conflict and violence in the Jordan basin, as Libiszewski argues, during the Arab–Israeli conflict 'water has been included in the dynamic of conflict mainly as an intervening variable, rather than as a catalyst in itself' (1995: 91–5). Similarly, Wolf (1995: 80) affirms that 'water resources were not a factor in Israeli strategic planning in the hostilities of 1967, 1978 or 1982 . . . the decision to go to war, and strategic decisions made during the fighting (including which territory it was necessary to capture), were not influenced by water security or the location of water resources'.

History and experience in this water-scarce basin demonstrate that sustainable solutions to water problems, whether domestic or international, always require cooperation, equitable sharing, and efficient utilisation among involved parties (Naff 1993: 7). This reinforces our main hypothesis that the very importance of water makes cooperation over water more likely than conflict. As in ancient times, the shared need for optimum management of this scarce resource will become a source of regional unity rather than regional discord. Indeed, as Dellapenna (1995) rightly suggests, water can become the key to building peace in the region if the two sides are prepared to exploit this possibility actively and effectively. Wolf makes the same point; 'the inextricable link between water and politics can be harnessed to help induce ever-increasing co-operation in planning or projects between otherwise hostile riparians, in essence 'leading' peace talks' (1995:3).

However, the complex environmental and hydrological issues cannot be resolved by political formulas negotiated by diplomats alone. The tasks of environmental rehabilitation and particularly of water resource development and efficient utilisation require a considerable investment of capital. Whatever the cost, however, it is certain to be much less than that of the only possible alternative, which is the collapse of the socio-economic foundation of all communities who live in a hydro-geological basin (Hillel 1994: 19). This is what the new generation of Jews who are born and bred in the environmental and climatological conditions of the region have realised. The fact is that any basin-wide and cooperative solution, technical and otherwise, depends on a fair resolution of the Israeli–Palestinian conflict in accordance with international law, whose cornerstone is the assumption that the allocation of scarce resources requires legal means, rather than coercive force. While to some sceptics such a basin-wide cooperation to achieve equitable solutions for such a vital issue seems unlikely, we argue that the alternatives are even more unlikely. Moreover, as we shall see in the next chapter, in the Euphrates–Tigris basin, with very different circumstances, the same conclusion is reached.

Notes

1 Zionism is an organised movement of world Jewry that arose in Europe in the late nineteenth century with the aim of constituting a Jewish state in Palestine. Modern Zionism is concerned with the development and support of the state of Israel.
2 For such evidence, see: Allan 1996; Allan with Court 1996; Allan and Karshenas 1996; Allan and Nicol 1996; Allan and Radwan 1996; Ben-Shahar *et al.* 1989; Feitelson 1996; Isaac and Shuval 1994; Kally and Fishelson 1993; Murakami *et al.* 1995; Peres with Naor 1993; Shuval 1992.
3 For a historical review of the climatic pattern of the Jordan river basin and its impact on the socio-economic, political, and religious aspects of the society, and to see how ancient residents of the region adapted to a water scarce environment by their choice of settlement sites and subsistence strategies, see Ngan (1991).
4 Although the Jordan River is the most extensively studied river system in the Middle East, there are wide discrepancies in data concerning its hydrological and topographical features. However, since these data are of secondary importance to this study, we do not reiterate them here. For detailed descriptions of the physiography and hydrological features of the Jordan River basin see among others: Bakour and Kolars 1994: 130–2; Beschorner 1992: 8–10; Kliot 1994: 175–85; Libiszewski 1995: 1–16; Lowi 1993a: 20–32; Murakami 1995: 69–79, Murakami and Musiake 1994: 117–27; Naff and Matson 1984: 17–23.
5 *The Register of International Rivers* (UN 1978: 15) indicates that the drainage area of the Jordan River system is divided between the Kingdom of Jordan (53.9 per cent), Syria (29.6 per cent), Israel and Palestinian territories (10.4 per cent), and Lebanon (6.1 per cent). However, some authors have other calculations. For instance, Libiszewski (1995: 3) believes that Jordan, Israel, and Palestinian territories cover about 80 per cent of the drainage area of the Jordan River system, while the figure presented by Bulloch and Darwish (1993: 40) is 64.5 per cent.
6 It is obvious that when the Zionists in the nineteenth century proposed a state stretching from the Nile to the Euphrates, the choice of

The Jordan River Basin 115

those borders was not based solely on historical considerations. The hydraulic orientation of the Zionist ideology is well reflected in their choice of borders. It is interesting to note that a recent survey by Ozanne (1996: 4) revealed that 'not even the right-wing Golan residents believe the Heights are an intrinsic religious part of the biblical land of Israel'.

7 Chaim Weizmann, the Zionist leader who eventually became Israel's first president, raised this issue in his letter to Lord Curzon, then British Foreign Secretary, following the San-Remo accord. He said:

> The accord draft France proposed not only separates Palestine from the Litani River, but also deprives Palestine from the Jordan River sources, the east coast of the Sea of Galilee and all the Yarmouk valley north of the Sykes–Picot line. I am quite sure you are aware of the expected bad future the Jewish national home would face when that proposal is carried out. You also know the great importance of the Litani River, Jordan River with its tributaries, and the Yarmouk River for Palestine. (Weisgall 1977)

8 The following discussion is primarily based on the persuasive analyses by two Jewish scholars from the Hebrew University of Jerusalem, Avner de-Shalit (1995) and Eran Feitelson (1996).

9 The Jewish National Fund (JNF) was one of the main organisations which was founded in 1901 to purchase land for the Jewish people in Palestine. The total area bought by the JNF up to the establishment of the state of Israel (1948) was 240 000 acres (96 000 hectares), most of which had to be reclaimed to be arable. (Israel Information Service 1996).

10 The Jewish National Fund (JNF), the institution responsible for forests and afforestation in Israel, describes this attitude as follows: 'In the beginning of the renewed settlement in the land of Israel, the new immigrants found a ruined and empty country, without a tree or a shadow. Therefore the first immigrants were determined to bring back to this country the green vegetation which had been stolen from it in previous generations, and about which we can learn from the Bible.' However, recent findings reveal that this is a myth, since in biblical times the country was not covered with trees (Felix, quoted in de-Shalit 1995: 85).

11 Most of the new settlements were started by Nahl, the agricultural unit of the Israeli Defense Forces, and subsequently transferred to civilian authorities (Naff and Matson 1984: 33).

12 According to Feitelson (1996: 18), several subsidy mechanisms were introduced: first, a direct subsidy from the general budget to the water sector; second, an intersectoral subsidy, whereby the domestic sector pays higher fees than the agricultural sector; and third, an implicit subsidy through which all users face the same tariff regardless of place. Thus settlements in mountainous areas and in the south, where water production costs are higher, receive an implicit subsidy at the expense of regions for which water supply costs are lower. This mechanism is still in effect today.

5
Water Politics in the Euphrates–Tigris Basin

Introduction

From the mid-1980s, when some American officials such as John Cooley (1984) from the US State Department and Richard Armitage, former assistant secretary of defense for international security affairs, came to believe that 'the water crisis in the Middle East is worsening and adding an extra dimension to prospective war scenarios' (Starr 1991: 36), the US administration has commissioned several studies and reviews of possible future water conflicts that might call for American intervention, predictably the majority of them in the Middle East (see for example: Naff and Matson 1984; Starr and Stoll 1987; US Army Corps of Engineers 1991). Moreover, according to Starr (1991: 17) and Bulloch and Darwish (1993: 16), both the Central Intelligence Agency (CIA) and the Pentagon have made their own assessments of potential water wars. Next to the Jordan River basin, the flash-point where water wars are claimed to be most likely is the Euphrates–Tigris basin (Starr 1991: 31; Postel 1992: 80; Bulloch and Darwish 1993: 16, 24–25, 31–32; George Joffe 1993: 76, 83; Peres with Naor 1993: 127; Hillel 1994: 103, 110; Barham 1996: 3). In 1992, a Pentagon report identified a water war between Syria and Turkey as one of the first contingencies. Some writers, such as Daniel Hillel (1994: 110), endorse this idea, asserting that 'conflict over water will be more than likely between Syria and Iraq, or possibly between both and Turkey'. George Joffe (1993: 83) claims that virtually all the major river systems in the region are either objects of actual conflict or likely to become subject to conflict. He explains that 'the conflict may or may not be primarily over the access to, and use of, water. It will, however, eventually concentrate over these two issues.'

Other writers such as Postel (1992: 74), are more equivocal, believing that water issues in this basin 'will foster either an unprecedented degree of cooperation or a combustible level of conflict'. A small number of scholars are, however, cautiously optimistic. Gleick (1994a: 6), for instance, argues that 'the need to manage jointly the shared water resources of the region may provide an unprecedented opportunity to move toward an era of cooperation and peace.' Given the growing thirst of the southern lands and the increasing need for cooperation among all the countries involved, Kolars and Mitchell (1991: 297) even suggest that, 'it is quite conceivable that we will see in our lifetime a *Pax Aquarum* in this part of the Middle East, one in which Turkey will play an important role'.

In this chapter, supporting the views of Gleick, Kolars and Mitchell, we shall argue that, for four reasons, the conflictual representation of hydropolitics in the Euphrates–Tigris basin is unconvincing. First, it attributes a degree of strategic significance to water which it does not essentially merit. Second, it overlooks the fact that in the Euphrates–Tigris basin, in marked contrast to the Jordan River basin, none of the riparian countries is facing an imminent water shortage. Third, given the current extensive irrigation and hydroelectric power projects in all the riparian countries and their unfolding development plans, it ignores the fact that the major water problems arise primarily from the mode of water allocation *within* states rather than to water allocation *between* states. Fourth, it turns a blind eye to the efforts which have been made by the riparian countries to avoid water conflict. Indeed, given that the region was a scene for the playing out of East–West intense rivalries, and given the territorial disputes and the general atmosphere of tension in the region, it is remarkable that the three parties have been engaged in continuous technical consultations and political negotiations over water since the early 1960s. This progressive process, as we shall see, has effectively undermined the likelihood of conflict between the riparian states.

Moreover, by an analysis of the relationship between water availability, as a cardinal environmental factor, and the rise of early civilisations in the Euphrates–Tigris basin, this chapter demonstrates that, contrary to what is conventionally believed, cooperation over control and appropriation of water resources has been the prevalent feature of water management policy in this basin from antiquity – a feature which characterised them as hydraulic civilisations (see Chapter 1). To develop this argument, we shall show that these civilisations had not only the engineering skills necessary for an efficient water-management system, but also the social experience and legal institutions required for maintaining the functionality of such a system. Hence, although water has sometimes been used as a defensive barrier or a destructive weapon, *water-imperative* conflicts have never occurred in this basin.

The second of the four reasons outlined above is the decisive one. The fact is that, despite some sensational political rhetoric and alarmist predictions, water scarcity in the Euphrates–Tigris basin has never been as acute as it has been in other water basins of the Middle East. Although uneven precipitation, recurrence of serious floods and droughts, and the extreme fluctuations of the twin rivers have always been a central concern of Mesopotamia's inhabitants, the peoples in this region have successfully come to terms with their natural environment. This is why, in sharp contrast to the contemporary unsustainable developments in the Jordan River basin and the Arabian Peninsula (see Chapters 4 and 6), projects implemented in the Euphrates–Tigris basin have not resulted in significant adverse ecological consequences and the real demand for water has not yet exceeded supply.

Of course, there have been acute tensions among the riparian countries, but we will argue that they are due to a host of different variables, including the East–West rivalry that grafted itself on to the political resentments and animosities which already existed in the region, rather than a consequence of water scarcity. On the contrary, we believe that the marginality of water scarcity in the basin has provided favourable circumstances in which the parties can use the water issue as a bargaining chip in their negotiations, and in due course as political leverage, without frustrating each other too much. This explains why downstream states in the basin do not resort to military force to prevent upstream developments, as long as water is the only bone of contention. In examining the latest developments in the region concerning the water diplomacy of Turkey, Syria, and Iraq we shall show how these countries are moving toward more cooperative arrangements, rather than towards more conflictual relations, as is conventionally believed. But first, we must outline the hydrological features of the Euphrates–Tigris basin.

Climatic and hydropolitical features

As mentioned earlier, the Middle East is a transitional zone between equatorial and middle latitude climates. It is generally characterised by aridity except for its northern belt, which is a well-watered area with a rugged landscape of high mountains and deep gorges where freezing temperatures prevail in the winter and much of the precipitation falls in the form of snow. Stretching from Turkey to Iran, this northern belt consists of two mountain ranges – the Pontus along the Black Sea and the Taurus along the Mediterranean. These mountains surround the Anatolian Plateau of Central Turkey and combine in a relatively cool and humid zone of Eastern Turkey, where both the Euphrates and Tigris Rivers originate only 30 km from each other. These headwater districts, enriched by autumn and spring rains and sustained by winter snows, nourish the two rivers which run separately onto the wide, flat, hot and progressively drier plain of Mesopotamia – the 'cradle of civilisation' where prehistoric humans are believed to have first turned to agriculture. The mountain belt divides again in Iran. Several important tributaries of the Tigris drain the Zagros mountains, extending southeast from the Turkish frontier to the Strait of Hormuz along the western edge of the Central Iranian plateau (Grolier 1993).

Thanks to its snowy mountains and abundant precipitation, Turkey is the richest country in the Middle East in terms of both ground and surface water resources. In addition to the catchment basins of the Euphrates and the Tigris, the catchments of Asiatic Turkey that flow to the Black Sea and to the Mediterranean are basically self-contained, with such a surplus flow that proposals to transfer water across the country into the Euphrates–Tigris basin or to the other parched areas of the Middle East have been suggested (Duma 1988; deShazo and Sutherlin 1994). According to data obtained from Turkish sources, average annual precipitation in Turkey is 501 bcm but, owing to excessive infiltration, transpiration and evaporation, only 186.05 bcm ends up as surface runoff. Considering present technological, topographical, and geological constraints, an estimated 95 bcm/yr of Turkey's surface water runoff plus an additional 11.6 bcm of ground water can be used. So far, however, only 30 bcm of these resources have been annually appropriated by Turkey (Tomanbay 1993: 53).

In contrast to Turkey's relative water abundance, more than half of Syrian and almost two-thirds of Iraqi territories are desert, with

Map 5.1 The Euphrates–Tigris basin

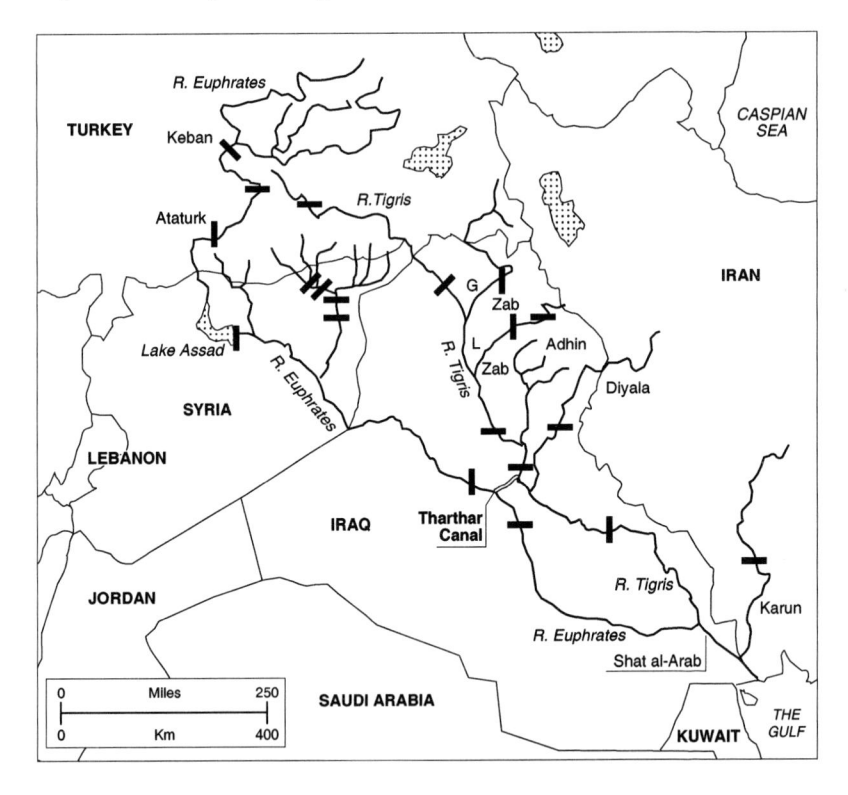

less than 250mm of rainfall per year, which is the minimum amount needed for rain-fed agriculture. Although Syria has other water resources, the Euphrates is the only major river crossing its territory with reliable annual flows. Accordingly, Syria depends heavily on the Euphrates, whose waters account for as much as 86 per cent of the water available to the country (Lowi 1995). In the case of Iraq, except for its mountainous north, which enjoys a milder climate and considerable rainfall, the climate is characterised by hot summers and cool winters. In the Mesopotamian Plain itself, the annual rainfall is meagre and the summer months are exceedingly hot and dry–median temperatures approach 50 degrees Celsius and daytime relative humidity is as low as 15 per cent.

Consequently the evaporative rate is very high, and agricultural production is totally reliant upon irrigation (Hillel 1994: 97; Lowi

1995). Although, like Syria, Iraq is heavily reliant upon the Euphrates waters, it is fortunate in having an alternative source of water in the Tigris River, whose headwaters are sligthly tapped (Starr 1991: 30; Gleick 1994a: 12).

The Euphrates–Tigris basin drains an area of around 900 000 km². From its headwaters to the sea, the Euphrates traverses a distance of 2700 km, covering an area of 444 000 km² which includes surface tributaries, wadis, and areas of purely subsurface recharge. Its twin, the Tigris, has a total length of 1900 km and embraces an overall area of 470 000 km², though its effective catchment area is estimated to be between 217 843 and 373 000 km² (Ewan Anderson 1991b: 12–3; Grolier 1993; Hillel 1994: 92). Regarding the basin's hydrological configuration, there are considerable discrepancies between data obtained from the riparian countries as well as figures suggested by foreign observers (see Tables 4.1, 5.2, 5.3).[1] For instance, while Naff and Matson (1984: 83), Starr and Stoll (1988: 11), Ewan Anderson (1991b: 12), and Lowi (1995) estimate that 28 per cent of the Euphrates basin is located in Turkish territory, 17 per cent in Syria, 40 per cent in Iraq, and 15 per cent in Saudi Arabia, Hillel (1994: 92) has a notably different estimate, claiming that some 40 per cent of the Euphrates basin is in the modern state of Turkey, 25 per cent in Syria, and 35 per cent in Iraq, and Pope (1995: 4) suggests that Iraqi territory accounts for 46 per cent of the Euphrates–Tigris drainage basin.

According to various sources, the Euphrates' annual flow is effectively generated within Turkey and Syria in the proportion of 88–90 per cent and 10–12 per cent respectively. But Beschorner (1992: 29) believes that over 98 per cent of Euphrates flow originates in Turkey. As for the Tigris basin, Ewan Anderson's (1991b: 13) assessment indicates that 12 per cent of it lies in Turkey, 0.2 per cent in Syria, 54 per cent in Iraq, and the rest in Iran. The figures produced by Hillel (1994: 92), however, suggest that about 20 per cent of the basin lies in Turkey, 78 per cent in Iraq, and only 2 per cent touch the northeastern corner of Syria.

The average annual flow of the two rivers is also a matter of controversy. Naff and Matson (1984: 86), Starr and Stoll (1988: 11), and Hillel (1994: 102) calculated that it is about 70–74 bcm/yr – the mean annual discharge of the Euphrates is estimated at approximately 32 bcm, and that of the Tigris 42 bcm. Gleick (1994a: 12) claims that the total flow of the two rivers exceeds 80 bcm/yr, of which about 33 bcm are generated in the Euphrates and 47 bcm in the Tigris. Hillel (1994: 95–6) and Murakami (1995: 35) claim that the mean natural flow of the Euphrates is about 30 bcm/yr at its entrance into Syria from Turkey, and 32 bcm/yr on leaving Syria, after having taken in two tributaries – the Balikh and the Khabur. But Barham (1994: 8) maintains that before the first dams were completed, the Euphrates carried 27 bcm of water across the Syrian border every year. As for the Tigris, Beschorner (1992: 29) claims that the average annual discharge of the river at the Turkish border is 16.8 bcm. But Hillel (1994: 95–6) maintains that where it enters Iraq its mean annual flow is between 20 and 23 bcm/yr, collecting an additional 25 to 29 bcm/yr in Iraq from its left-bank tributaries. Relying on figures suggested by Beaumont et al. (1988: 84), George Joffe (1993: 73) claims that 'the headwaters of both the Tigris and the Euphrates generate a water supply of around 40 bcm annually. ... Both rivers provide Iraq with around 50 bcm annually before uniting into the Shat al-Arab at al-Qurna, with the Euphrates providing 57.5 per cent of the flow and the Tigris the balance'.

A feature common to both rivers is that their discharge is subject to extreme fluctuations, both seasonally and annually. The rivers' annual flow cycle in natural conditions have three distinct periods: first, a minor rise in the rainy season, from November to March, during which 20 per cent of the annual flow occurs; second, a major rise in spring, during which the rivers are in spate as a result of snowmelt in the mountains of Turkey augmented by seasonal rainfall. This period normally lasts from April to June and contributes 70 per cent of the total annual flow. Third, a summer low water, lasting from July to October, with only 10 per cent of annual flow (Hardan 1993: 76; Hillel 1994: 97). The ebb and flow in the Euphrates–Tigris river system is so extreme that during the second phase sometimes the rivers' flux may be 10 times as great as that of the low-flow period of late summer and half of the annual discharge is generated during April and May (Gleick 1994a: 12; Hillel 1994: 97). The minimum flow of the Euphrates is reported to be as low as 100 to 180 cubic meters per second (cm/s) during the summer, while the maximum flow could reach 5200 to 7000 cm/s when the spring snows melt (Bulloch and Darwish 1993: 68–9; Gleick 1994a: 12). Hardan (1993: 75) declares that the maximum recorded flow of the Euphrates at Hit was 7390 cm/s on 13 May 1969 and the minimum flow was 50 cm/s in July 1974 when both Turkey and Syria simultaneously decided to fill their huge, newly built reservoirs of Keban and Tabqa, respectively.

Besides the extreme seasonal variation in the velocity of the rivers, the annual discharge varies greatly from year to year. According to Beschorner (1992: 29), the lowest annual flow ever measured in Euphrates was 16 bcm, and the highest was 50 bcm. But Gleick (1994a: 12) claims that in dry years, runoff in the Euphrates has amounted to as little as 30 per cent of the annual average flow. Data recorded from 1924 to 1973 shows that the mean annual discharge of the Euphrates river at Hit, in Iraq, varies from 10.6 to 63.4 bcm/yr (see Kolars and Mitchell 1991: 94–6). Available records indicate that the Tigris also exhibits considerable annual variations: from a minimum of 16.9 to a maximum of 58.7 bcm (Hillel 1994: 102).

As far as agriculture is concerned, the rivers' seasonal fluctuation is very unfavourable – its rise is too late for winter crops and too early for summer crops. Therefore, water must be diverted from the rivers during the period of high crop demand and low water flow. This means that considerable engineering capacity and an efficient water-management system are required to control the rivers' flow effectively. Since both rivers are subject to major fluctuations in flow, such a system is needed both to check the troublesome inundation of the land by floods and to preserve the floodwaters for irrigation during the main growing season. Thus, although one function of upstream dam-building, water abstraction, may cause controversy, another, that of discharge control, is not only beneficial but necessary, to the downstream state.

Another feature common to both rivers is the heavy concentration of suspended sediment in their waters especially at the time of seasonal floods. According to Hillel (1994: 93), during the flood season these rivers carry 'as much as 3 million tons of eroded soil from the highlands in a single day' most of which settles en route and provides the rich deposit of alluvium that fills the Mesopotamian Plain. As Hillel points out, little of this sediment reaches the sea. Such a characteristic adversely affects the storage capacity of artificial reservoirs constructed by the riparian states, particularly in the upstream area. Most analysts simply ignore the long-term repercussions of this significant ecological fact when they talk about the ability of upstream states to starve out downstream neighbours by cutting off their water supply (see for instance Lowi 1995, Falkenmark 1986a).

In the next section we shall discuss the dynamics of water-management policies in the Euphrates–Tigris basin to clarify that, first, contrary to conventional understanding, water has never been a source of strategic rivalry in this basin: indeed the rise of great hydraulic civilisations which flourished in the countries of this basin reflected their extensive and systematic cooperation over the control and distribution of water resources. Second, although the engineering approach to water management has been the dominant approach in all riparian communities along the two rivers, this has not led to exhaustion of natural water resources, as it did in the Jordan River basin and in the Arabian Peninsula; indeed there has always been an overall surplus of water in the Euphrates–Tigris basin.

Evolution of water management in Mesopotamia

Ancient Mesopotamia

Since 6000 BCE, successive civilisations that settled and flourished in Mesopotamia[2] have had an engineering approach to water resources management – which is essentially concerned with supply augmentation and satisfaction of demand for water rather than demand management (see Chapter 2). The first recorded intensive human manipulation of water supplies for agriculture and domestic purposes is linked to the domestication of crop plants, particularly grains, in Mesopotamia (Bakour and Kolars 1994: 125). Because of the climatic and hydrological conditions discussed above, the inhabitants of Mesopotamia had to resort to irrigated farming, an enterprise which required a highly organised and cooperative structure to mobilise, coordinate, and direct the efforts of the workers. The basic contributing factor in the development of civilisation was the seasonal deposit of sedimentation by silt-saturated floodwaters below the delivery point of tributary streams from the highlands. Such perennial flood areas had to be cleared of debris, penetrated by drainage canals to eliminate marshland and swamp, and protected by artificial levees to control excessive water and to facilitate the development of irrigation. At the early stages of settlement, villagers learned to cooperate to build primitive irrigation canals and ditches. Subsequently, elaboration of irrigation technology – especially the invention of the wheel – and modification of land use and cultivation patterns led to the creation and growth of city states that, according to Robert Adams (1965, ch. 4), by 4000 to 3000 BCE were able to muster the skills and labour to build major irrigation and flood control networks. These units were eventually merged into governmental centres – kingdoms and empires – which could define and enforce land claims and water rights and protect access to such facilities. As engineers in Mesopotamia found new ways of managing and controlling the natural force of water, civilisation moved further away from the natural course of water resources. Nevertheless, the social, economic, and political structure of these agrarian civilisations and their viability and development relied upon a well-developed and well-maintained hydro-agricultural system. Indeed, the challange to make efficient use of water resources continued to play an important role in the emergence of large-scale infrastructures and organised institutions in the successive ancient civilisations of the Sumerian, Babylonian, Assyrian, and Aramean Kingdoms. They all organised a system of flood control, developed complex irrigation networks, and raised considerable power based on a hydraulic economy. They also used the rivers for navigation, transportation, and urban development.[3] Regarding the troublesome seasonal floods of the twin rivers, maintaining an efficient water management system not only demanded considerable engineering knowledge and skill, but also necessitated a specific type of division of labour and large-scale cooperation. These needs were reflected in the Code of Hammurabi, which established water laws and rules for maintaining irrigation projects (Hardan 1993: 73; Hatami and Gleick 1994: 10).

Significantly, the collapse of these civilisations was associated with deterioration in their hydraulic projects and agricultural activities, induced either by natural or human causes or both. Successive periods of droughts and floods resulted in frequent change of the rivers' courses and sedimentation of irrigation networks. Historical accounts and the remains of hydraulic structure such as water wheels, spurs, dikes, and towering heaps, which are still visible in most parts of the Mesopotamian Plain, indicate that 'canals silted up if neglected for only a couple of months, and that operation of irrigation systems required tremendous cleaning and maintenance' (Hardan 1993: 73).

Mesopotamia was a land frequently subject to foreign invasion and political domination of successive conquerors. This process continued from antiquity until the First World War, during which British troops took the area. The need for self-defence led the ancient Mesopotamians to build canals around their walled settlements. The cities were, thus, both supplied and defended by hydraulic projects. King Nebuchadnezzar of Babylon (605 to 562 BCE), for instance, built immense walls around the city and used the Euphrates River and a series of canals as defensive moats. He is quoted as follows:

> To strengthen the defenses of Babylon, I had a mighty dike of earth thrown up, above the other, from the banks of the Tigris to that of the Euphrates ... and I surrounded the city with a great expanse of water, with waves on it like the sea. (Drower cited in Gleick 1993c: 109)

The irony is that, while water has had a highly significant role in the establishment and development of numerous civilisations in Mesopotamia, it has also been used as the most destructive weapon in the hands of enemies who wanted to conquer these civilisations. During these recurrent attacks and counter-attacks, depending on environmental conditions, water and water distribution systems were treated as strategic military goals, targets of military attack, or as weapons.

Having a secure access to drinking water or depriving the enemy of it was always a strategic military goal especially in the desert areas. Assyrian forces adopted this strategy in their battle against Elam and Arabia, seizing or drying up their water wells. In their invasion of Babylon in 689 BCE, they attacked the water supply canals to the city as a military target whose destruction would bring the enemy to its knees. According to Herodotus, Cyrus successfully invaded Babylon by diverting the Euphrates above the city into the desert and marching his troops into the city along the dry riverbed (Hatami and Gleick 1993: 5).

Such event have led some writers to the conclusion that 'history reveals that water has frequently provided a *justification for going to war*' and 'no region has seen more *water-related conflicts* that the Middle East', where 'the *scarcity of water* has played a central role in defining the political relationships of the region for thousands of years'. '*Competition* for the limited water resources of the area is not new; people have been *fighting over*, and with, water since ancient times.' (Gleick 1994a: 6–8 [emphasis added]). In order to substantate these claims, Hatami and Gleick (1993) have examined myths, legends, and historical accounts of the region and compiled a chronology of such so-called water conflicts in the ancient Middle East, most of which occurred in the valleys of Euphrates–Tigris basin (see Table 5.1). A careful scrutiny of those events, however, discloses that *none of them* has been a *water conflict* in the proper sense of the term.

Although it is difficult to disentangle the many intertwined causes of conflict among nations, in the cases cited in Table 5.1, it is evident that competition or dispute over water as a scarce resource has rarely, if ever, been the cause of violent conflict in Mesopotamia. On the contrary, it has been the abundance of water which has enabled people to benefit from it as the bedrock of their ecomony and to use it as a defensive barrier or offensive weapon. The desire of kingdoms and empires, either those who arose in this area or those who invaded it, for expansion of their territories or subjugation of others, is very different from their going to war to capture *scarce* water resources in a specific area. There is also a sharp difference between conflict over a shared water body and using water as a military target or as a weapon against an enemy.

Hatami and Glieck claim to identify examples of genuine 'water-imperative' military conflicts between parties who had shared interests in a scarce water body. However, the examples cited by Hatami and Glieck (see Table 5.1) are cases in which water is used as an *instrument* for the punishment of humanity for its sins (by a deity or god); or as an offensive weapon (by Lagash, Esarhaddon, Assurbanipal, Nebopolassar, and Cyrus); or a defensive weapon (by Abi-Eshuh, Moses, Hezekiah, and Nebuchadnezzar); or as a target of destruction with the intention of eliminating the enemy's ability to recover (by Sargon II, Sennacherib, and Alexander). But none of these represents a genuine 'water-generated' conflict.

Indeed, history reveals that the justification for going to war *has invariably been something other than water*; people have fought *with* water since ancient times but not *over* water. What is remarkable to note is the major role that water has played in the creation, configuration, and development of civilisation in this region since antiquity. Hence, since water was the substructure of civilisation in the Mesopotamian Plain, and since the construction, operation, and development of water distribution systems required an organised and cooperative structure to mobilise and direct the collective efforts for a collective benefit, cooperation, not conflict, was the prevailing mode of conduct. The Code of Hammurabi, which, in the second millennium BCE, established laws and rules for maintaining an efficient water management system, indicates that the Mesopotamian civilisations not only had considerable hydro-engineering knowledge and skill, but also had well-established civil institutions to handle the system so well that at its peak it sustained some 20 million inhabitants (see: Kliot 1994: 117).

Table 5.1 A partial chronology of conflict over water in the ancient Middle East

BCE The Flood
An ancient Sumerian legend recounts the deeds of the deity Ea [god of water], who punishes humanity for its sins by inflicting the Earth with a six-day storm. The Sumerian myth parallels the biblical account of Noah and the Deluge, although some details differ.

2500 BCE Lagash–Umma border dispute
The dispute over the 'Gu'edena' (edge of paradise) region begins. Urlama, King of Lagash from 2450 to 2400 BCE, diverts water from this region to boundary canals, drying up boundary ditches to deprive Umma of water. His son Urlama II cuts off the water supply to Girsu, a city in Umma.

1820–1684 BCE Abi-Eshuh v. Iluma-Ilum
Abi-Eshuh, king of Babylon, dams the Tigris to prevent the retreat of rebels led by Iluma-Ilum, who declared the independence of Babylon. This failed attempt marks the decline of the Sumerians, who had reached their apex under Hammurabi.

1200 BCE Moses and the parting of the Red Sea
When Moses and the retreating Jews find themselves trapped between the pharaoh's army and the Red Sea, Moses miraculously parts the waters of the Red Sea, allowing his followers to escape. The waters close behind them and cut off the Egyptians.

720–705 BCE Sargon II Destroys Armenian waterworks
After a successful campaign against the Haldians of Armenia, Sargon II of Assyria destroys their intricate irrigation network and floods their land.

705–682 BCE Sennacherib and the Fall of Babylon
In quelling rebellious Assyrians in 695 BCE, Sennacherib razes Babylon and diverts one of the principal irrigation canals so that its waters wash over the ruins.

Sennacherib and Hezekiah
As recounted in Chronicles 32:3, Hezekiah digs a well outside the walls of Jerusalem and uses a conduit to bring in water. Preparing for a possible siege by Sennacherib, he cuts off water supplies outside of the city walls, and Jerusalem survives the attack.

681–669 BCE Esarhaddon and the siege of Tyre
Esarhaddon, an Assyrian, refers to an earlier period when gods, angered by insolent mortals, create a destructive flood. According to inscriptions recorded during his reign, Esarhaddon besieges Tyre, cutting off food and water.

669–29 BCE Assurbanipal, siege of Tyre, drying of wells
Assurbanipal's inscriptions also refer to a siege against Tyre, although scholars attribute it to Esarhaddon. In campaigns against both Arabia and Elam in 645 BCE, Assurbanipal, son of Esarhaddon, dries up wells to deprive Elamite troops. He also guards wells from Arabian fugitives in an earlier Arabian war. On his return from victorious battle against Elam, Assurbanipal floods the city of Sapibel, an ally of Elam. According

to inscriptions, he dams the Ulai River with the bodies of dead Elamite soldiers and deprives dead Elamite kings of their food and water offerings.

612 BCE Fall of Ninevah in Assyria

A coalition of Egyptian, Median (Persian) and Babylonian forces attacks and destroys Ninevah, the capital of Assyria. Nebuchadnezzar's father, Nebopolassar, leads the Babylonians. The converging armies divert the Khosr River to create a flood, which allows them to elevate their siege engines on rafts.

605–562 BCE Nebuchadnezzar uses water to defend Babylon

Nebuchadnezzar builds immense walls around Babylon, using the Euphrates and canals as defensive moats surrounding the inner castle.

558–528 BCE Cyrus the Great digs 360 canals

On his way from Sardis to defeat Nabonidus at Babylon, Cyrus faces a powerful tributary of the Tigris, probably the Diyalah. According to Herodotus's account, the river drowns his royal white horse and presents a formidable obstacle to his march. Cyrus, angered by the 'insolence' of the river, halts his army and orders them to cut 360 canals to divert the river's flow. Other historians argue that Cyrus needed the water to maintain his troops on their southward journey, while another asserts that the construction was an attempt to win the confidence of the locals.

539 BCE Cyrus the Great invades Babylon

According to Herodotus, Cyrus invades Babylon by diverting the Euphrates above the city and marching troops along the dry riverbed.

355–323 BCE Alexander the Great destroys Persian dams

Returning from the razing of Persepolis, Alexander proceeds to India. After the Indian campaigns, he heads back to Babylon via the Persian Gulf and Tigris, where he tears down defensive weirs that the Persians had constructed along the river. Arrian describes Alexander's disdain for the Persians' attempt to block navigation, which he saw as 'unbecoming to men who are victorious in battle'.

Source: Hatami and Gleick (1993; 1994: 11)

The people who successively conquered this region often enriched the Mesopotamian culture as they introduced their traditional beliefs, practices, and customs. However, because of environmental conditions peculiar to Mesopotamia and the existence of an already complex civilisation there, these newcomers eventually adopted the indigenous culture, enhanced the inherited technical achievements in irrigation and agriculture, and advanced the social and civil institutions proper to a hydraulic civilisation. The heritage passed through successive generations and empires and flourished in the

Arab–Islamic era, especially during the time of the Abbasid caliphs (750–1258 CE). In this era, Mesopotamia was a significant part of the Arab Empire in which agricultural development was marked by adaption of water-intensive plants such as cotton, rice, and sorghum (Watson 1983). This led to the intensification of irrigation and to a wide extension of large-scale irrigation systems, accompanied by a high level of crop production. This dazzling era was brought to an end, however, when Mesopotamia was devastated by Mongol hordes. Ironically, the role of water as a destructive weapon was decisive. Flooding the camp of the last defending Abbasid army by breaching the dikes of the Tigris River, Mongol invaders routed the Abbasid caliphate in 1258 CE. They sacked the area and massacred the inhabitants, as they did in other centres of civilisation, and destroyed the extensive hydraulic substructure of the region. But this was not a conflict caused by water.

In the following centuries, Mesopotamia declined in political importance, cultural ascendancy, and prosperity. Even though during the reign of Ottoman Turks the hydraulic system was partially repaired and relative prosperity came back to the region, with the empire's decline and resultant political instability, many irrigation installations fell into ruin, canals silted up and desertification advanced.' According to Hardan (1993: 73), this condition prevailed in the region throughout the nineteenth century. During the First World War, the region experienced a new foreign invasion which led to the disintegration of the Ottoman Empire. Following the conclusion of the Treaty of Sèvres (1920), the geopolitical map of the entire Middle East drastically changed. Among its consequences was the division of the Euphrates–Tigris basin between three parties – Turkey, Syria, and Iraq. This event opened a new page in the history of Mesopotamia in which hydropolitics gradually became a major international issue.

Modern Mesopotamia

By the Treaty of Sèvres (1920), the Mesopotamian Plain was mainly divided between the new states of Iraq and Syria, which were under British and French Mandates from 1920 until their independence in 1932 and 1941 respectively, while the headwaters of the twin rivers remained in Turkish territory. Although the Euphrates and Tigris rivers attracted the attention of planners in the early years of the Turkish Republic, the remoteness of the region and more pressing matters facing the Turkish government prevented any major water projects at that time (Kolars and Mitchell 1991: 25). Moreover, the relative abundance of water in the basin which could satisfy the local needs of all parties facilitated a peaceful course of development in the region. Until 1975, neither the absence of integrated planning for the entire basin nor the lack of comprehensive agreement among three riparian countries on a water-sharing regime, led to any international friction among them. Even those who are not optimistic about the future of hydropolitics in the Middle East acknowledge that, because of the existence of a marked surplus of water in Mesopotamia, 'until the 1970s no overt or covert conflict had ever evolved around water usage of the Tigris–Euphrates' (Kliot 1994: 161) and 'the projects implemented in this period did not result in significant adverse ecological consequences' (Naff and Matson 1984: 92).

However, the relative abundance of water in the region did not prevent the parties from conducting some bilateral or trilateral negotiations. Since the disintegration of the Ottoman Empire, the three parties generally pursued a cooperative policy in order to establish principles pertaining to the right of downstream countries to waters entering their territory, and to determine the best mechanism for the optimal management of the two rivers. This cooperative mode of conduct has often been overlooked by Western commentators who incline towards conflict-orientated scenarios. It is, therefore, necessary to explain briefly certain international water-related agreements which have been arrived at since partition of the Euphrates–Tigris basin among the three parties. This will help us to identify the neglected regulatory framework established by these accords in relation to regional water management, to understand the legal position of riparian parties with regard to the issue of water sharing arrangements, and to recognise their perception of hydropolitics. This, in turn, will enable us to determine whether hydropolitics as a regional phenomenon has been a divisive or a cohesive issue.

Before looking at the water management policies adopted by the three riparian countries, it should be noted that there are two distinct phases in the growth of water resources development in modern Mesopotamia. In the first phase, from the collapse of the Ottoman Empire in 1918 to the 1960s, flood control was the main concern of riparian parties particularly of Iraq. In the second phase, from the 1960s until now, all riparian countries, especially Turkey, with the help of modern technology embarked on massive water development schemes including hydropower generation. It is in this second phase that Western strategists have repeatedly warned that the question of water is becoming a more explosive Middle East issue than that of oil (see Chapter 1). During the first phase, there were no signs of water conflict; a few sizeable hydrological projects which were undertaken in downstream areas concentrated on flood control and rehabilitation of ancient irrigation networks, and riparian countries harmoniously managed to tackle these water problems of flood control and irrigation, through negotiations and cooperative measures. The second phase is a period in which the development of the Euphrates by the riparian countries has considerably grown in significance, and it has been marked by a series of projects that are portrayed by most writers as exemplifying the lack of cooperation among the riparian states and manifesting 'the battle for water in the Middle East'.

The first phase (1918–60)

During this period, the dominant approach to water resources management was much the same as in ancient time – namely the engineering approach. Irrigation was the dominant use of water, mainly in the southern part of the basin, and due to underdevelopment much of the waters from both rivers drained in to the Gulf. Turkey, as the upstream country, until 1974 used only 3 per cent of the Euphrates water (Kliot 1994: 136) and the Tigris River is still sparsely utilised in that country. Syria was the second to exploit the water of the Euphrates in recent times. After the Second World War with the introduction of motor-pumps, Syria's irrigated agriculture rapidly developed. According to Syrian official sources, the irrigated area increased from 295 000 ha in 1945 to 657 000 ha in 1962 (Wakil 1993: 64).[4] Despite these developments, which led to full exploitation of all the river valley low plains suitable for irrigation, Syria was extracting only between 2.5 to 3 bcm of the Euphrates flow in the 1960s, a trend which carried on into the 1970s and 1980s (Beaumont 1978: 40; Kliot 1994: 139). Further expansion of irrigated agriculture required state agencies to construct huge reservoirs, large-scale irrigation networks, and massive projects of land reclamation in the following decades.

As far as Iraq is concerned, in both ancient and modern times, the management of both the Euphrates and the Tigris rivers had its beginnings in southern Mesopotamia – the site of the modern state of Iraq. River management in modern times began with the report of British hydrological engineer, William Wilcox, to the

Ottoman Empire in 1911. Among his suggestions was the Hindiya Barrage on the Euphrates, near Babylon, which was constructed between 1911 and 1914 to control water and provide it, all year round, for irrigation through reconstructed canals dating from ancient and medieval times.[5] Under the British Mandate the Department of Irrigation was established and a program for hydrological data collection and for the design and construction of irrigation projects was initiated. However, it was after the Second World War that the largest projects were implemented (Hardan 1993: 74).

During this period, the conclusion of treaties, conventions, and protocols between the riparian parties are clear indicators of the dominance of a cooperative mode of relations between them. The first international agreement relating to the utilisation of water in Mesopotamia is the Franco-British Convention of 1920, according to which the mandatory powers agreed to establish a committee to examine and coordinate the water utilisation of the Euphrates and Tigris.[6] This convention, which in accordance with the principle of succession of states with regard to treaties was inherited by Syria and Iraq, established the necessity for a mutual agreement before carrying out 'any irrigation project prepared by the French Government, ... the realisation of which would considerably diminish the waters of the Tigris and the Euphrates' (Chalabi and Majzoub 1995: 193).

There is also a Franco-Turkish agreement of 20 October 1921 which is concerned with the upstream utilisation of the waters of the Koveik River[7] by Turkey and the possibility of tapping the waters of the Euphrates to compensate for downstream users in Northern Syria – namely the Aleppo district. Although the Koveik does not contribute to the Euphrates river, it exemplifies the upstream–downstream water use problem which is now the main issue in the region. This agreement shows that the riparian parties have been trying to solve their water problems in such an amicable way as to give equal satisfaction to all parties – a point which alarmist analysts tend to overlook.

This explicit appreciation of the rights of downstream parties was also reflected in the Treaty of Lausanne (1923), which noted that 'Turkey should confer with Iraq before beginning any activities that may alter the flow of the Euphrates' (Bakour and Kolars 1994: 139). Subsequently, the Friendship and Neighbourly Relations Convention of 30 May 1926, signed in Ankara between Turkey and Syria under the French Mandate, maintained that 'to satisfy the needs of the regions at present irrigated by the waters of Koveik, and those of the town of Aleppo ... Turkey must increase the flow of the Koveik, or even, authorise ... an off-take of the waters of the Euphrates, or perhaps combine both methods' (Art. XII.1). Moreover, the French–Turkish Protocol of 3 May 1930, which commits the two parties to coordinating any plans to utilise the Euphrates waters, not only confirmed the agreements previously signed between France and Turkey, but also dealt with the questions raised by the joint ownership of the Tigris (Chalabi and Majzoub 1995: 193–4; Kliot 1994: 161).

After the Second World War, Iraq and Turkey began exchanging information on the subject of flood prevention. Soon it was recognised that the optimal water management mechanism for the Euphrates–Tigris rivers, with such an extreme seasonal and annual variance (see above), was for flood control and water storage measures to be implemented in upstream areas. Clearly, if either Turkey or Iraq had perceived water management to be a strategic issue, such a mutual understanding would have been impossible. A Treaty of Friendship and Good Neighbourly Relations and an attached Protocol concerning problems arising from the rivers running through their territories, was signed in March 1946 by the two states. According to these agreements, Turkey obliged itself to begin monitoring the two rivers and to share relevant data with Iraq. Moreover, Turkey not only agreed to report to Iraq on all its plans to utilise the Euphrates and Tigris but it also gave Iraq the right to construct dams within Turkish territory to improve the Euphrates water flow within Iraq (Kliot 1994: 161; Kolars 1994: 64). The essence of the Protocol indicates that both parties perceived the issue as a technical and engineering question which required collaboration between the upstream and downstream countries. Its principal objective was the construction of protection and observation posts on Turkish territory to prevent downstream flooding and, therefore, it mainly benefited Iraq. Accordingly, the cost of all installations, preliminary research works, and exchange of information were to be borne by Iraq, but the maintenance costs were to be shared by both parties (Chalabi and Majzoub 1995: 194).

In the 1950s, following the establishment of the Board of Development, the Ministry of Development, and the Ministry of Agrarian Reform in Iraq, a productive period of national planning arrived. With the help of foreign firms, a second barrage which was constructed near the city of Ramadi allowed Euphrates floodwaters to be diverted to Lake Habbaniyah and the Abu Dibis depression – with a total storage capacity of some 45 bcm – permitting a measure of flood control through three regulators, and three canals (Hardan 1993: 74; Hillel 1994: 99; Naff and Matson 1984: 89). Comprehensive integration of the national and regional programmes, however, was disrupted in 1958, when a leftist pan-Arab army coup established a republic in Iraq with a foreign policy orientated toward the USSR.

Although no comprehensive international treaty which could regulate the sharing or common exploitation of the rivers was concluded between the three riparian countries during this period, the aforementioned bilateral agreements established general principles concerning the right of downstream countries to waters entering their territory. As Chalabi and Majzoub (1995: 195) note, despite political disturbances in the region, these bilateral agreements generally worked well and could serve as a basis for a more extensive cooperation between the riparian countries. However, it must be admitted that cooperation over joint projects was still very limited in this period. Apart from regional political obstacles, this was mainly because, in marked contrast to the main Jordan River riparian states, the three riparians of the Euphrates and Tigris had plenty of water at their disposal. As mentioned above, until the mid-1970s Turkey used only about 3 per cent of the Euphrates water, Syria near 10 per cent, and Iraq slightly more than 50 per cent (Kliot 1994: 136, 144). Including the much less-developed water resources of the Tigris River, it is clear that the Euphrates–Tigris system had substantial water left after current demands were met in all riparian countries. In such circumstances none of the parties perceived a potential water crisis on the horizon which could have led them to negotiate a definitive settlement for water sharing on a regional scale.

The second phase (1960 onwards)

Although until the mid-1970s no international conflict had ever arisen over water usage in the Euphrates–Tigris basin, the water resource situation began to change when, in the 1960s, both Turkey and Syria started to draw up plans for large-scale exploitation of the Euphrates and, to a lesser extent, the Tigris. Iraq, the major user of water in the basin, perceived it as a menacing strategic challenge that both Turkey and Syria were determined to raise their claims on their shares. A series of trilateral and bilateral diplomatic discussions were held, in which each party put forth its projected demands; together, these exceeded the natural yield of the river. The failure of the parties during these negotiations to reach a formal and comprehensive water allocation agreement impelled each country to embark on its own development plans on the portions of the rivers in its territory. Since then, a strategic importance has been given to the water resources of the basin and, accordingly, a growing anxiety has been expressed by observers who claim that there would not be sufficient water to meet Iraq's needs if all upstream proposed schemes were implemented (Starr and Stoll 1988: 6; George Joffe 1993: 75; Gleick 1994a: 13).

Observers have repeatedly warned that since Turkey, Syria, and Iraq all have ambitious plans for exploiting the waters of the Euphrates, 'trouble may be around the corner' (Gowers and Walker 1989: 7). Starr (1991: 31) claims that continued stalemate and the unilateral construction of new dams could lead to 'escalating disputes and armed confrontation'. Others, such as Hillel (1994: 110) and Kliot (1994: 143), argue that, since Iraq may lose as much as 80 per cent of its Euphrates inflow owing to upstream water withdrawals, the likelihood of conflict in the area is very strong, either between Syria and Iraq or between either or both and Turkey. And although Sandra Postel of the World Resources Institute notes that 'the Tigris–Euphrates has substantial water left after current demands are met', 'Yet, ' she cautions, 'the failure of the basin's three countries – Iraq, Syria, and Turkey – to reach water-sharing agreements has created an atmosphere of competition and mistrust that could breed future conflict' (Postel 1992: 80).

The need for a basin-wide water-sharing agreement, but the failure of riparian countries to establish such an agreement, led observers such as Gleick to attribute the lack of cooperation in part to wider issues of international politics:

> For 30 years, negotiations over the Euphrates among Turkey, Syria, and Iraq have produced no lasting agreement, in part because the three countries have long been at odds with each other. For example, Syria and Iraq have opposed Turkey over its membership in NATO, and Syria and Turkey opposed Iraqi military actions in the 1970s. In the 1980s, Turkey and Iraq tended to band together against Syrian military aggression, and Turkey and Syria sided with the allied forces against Iraq during the Persian Gulf War in the early 1990s. (Gleick 1994a: 13)

The Euphrates–Tigris Basin 137

Lowi (1995) attributes the lack of agreement to territorial disputes and personal rivalry between the political leaderships of the riparian countries, saying that:

There has been a need for basin-wide agreement since the mid-1960s, when upstream use of the Euphrates flow began to put pressure on downstream consumption practices. There have been several unsuccessful attempts at promoting a tripartite accord, but in every instance, the tense political relations in the basin have impeded co-operative outcomes. Syria and Turkey have a simmering territorial dispute concerning Alexandretta (the Hatay province) that dates back to the Mandate period. In recent years, the two countries have been at odds over the Kurdish insurgency movement. Furthermore, the Ba'athi rulers in Syria and Iraq have been engaged in a highly acrimonious, personalistic conflict since 1968. (Lowi 1995: 136–8)[8]

Chalabi and Majzoub (1995) likewise believe the upsurge of hydropolitics to be due to political competition between the region's elites. They argue that:

Following the Suez Affair in 1965, Egyptian propaganda made so much publicity over the Aswan High Dam on the Nile that the authorities of other countries, which considered themselves as 'modern' or 'liberal', felt that to win over public opinion they also had to embark on their own high dam project. These countries went ahead without any mutual consultation as to their real water needs. (Chalabi and Majzoub 1995: 197)

Accusing all three riparian countries of the Euphrates–Tigris basin of imitating Cairo in a 'race for regional leadership', Chalabi and Majzoub claim that since 1960 'all co-operation was replaced by a mere quest for prestige. Each country tried to capture the collective imagination of its people, an indispensable factor in a still-lacking interior cohesion'. However, Chalabi and Majzoub believe that 'what is striking in the development of the Euphrates basin is not so much the local and traditional oppositions proper to the countries of the Middle East, but the transposition of the East–West rivalry for the economic and technical supremacy that both camps practiced with such relentless determination' (1995: 197). According to this analysis, what was critical was the East–West competition which 'came to graft itself on to the internal rivalries between the countries in the region.' It was specially in the 'East–West context' that the development of the Euphrates by the three riparian countries gained considerable strategic significance; and for 'mere political reasons' the contracts were signed for the construction of separate hydraulic projects in three riparian countries, a process which has been referred to as 'the battle for water in the Middle East' (Ibid.: 198).

The views of these observers, whatever their differences, indicate that the so-called water conflict in the Euphrates–Tigris basin is not an autonomous issue, but is the result of numerous causes which have nothing to do with 'water scarcity' in the proper sense. If these countries are at odds with each other for reasons other than water scarcity, such as territorial disputes or specific internal problems, which can be aggravated or ameliorated by external influences, or if the region is divided into competitive domains which are subordinate to antagonistic superpowers in a bipolar or multipolar world order, we cannot argue that the lack of cooperation between the riparian countries in their hydropolitics policies is a direct result of apparent or potential water shortages in the region. The only genuine case of water conflict in modern Mesopotamia – the Syrian–Iraqi crisis of spring 1975 – readily receded as a result of international mediation and it has not reappeared as a critical issue which could antagonise relations between the two riparian states. This is particularly remarkable in view of the icy relations between Syria and Iraq in the last three decades. Moreover, even if we attribute the *principal* role to water shortage in that 1975 conflict, an attribution which is contestable, it was at worst a very temporary condition caused by the simultaneous filling of the gigantic reservoirs of Keban and Tabqa dam in Turkey and Syria in a drought year. The bilateral agreements and trilateral consultations among the riparian countries and their generally conciliatory approaches to water issues have prevented such an incident from happening again and are likely to do so in the future.

Looking at the issue from a legal point of view, George Joffe suggests that 'the problem of water division between Turkey, Syria and Iraq is complicated by the fact that *there are no guidelines in international law as to how this should be done*' [emphasis added], claiming that, 'without a regionally acceptable cooperative water regime being instituted in the near future, conflict over water use in the Tigris–Euphrates system seems inevitable' (1993: 76). But it appears that Joffe has overlooked both the Convention on the Protection and Use of Transboundary Watercourses and International Lakes'[9] and more than two decades of endeavour and deliberation by the International Law Commission for the development and codification of the non-navigational uses of international watercourses (ILC 1991).[10] Both of these significant international legal achievements lay emphasis, among suggested general principles, upon equitable utilisation and the obligation not to cause significant harm to other riparian parties.

However, Hillel (1994: 10–3) argues that the claims and counter-claims of the three riparian states are so complex that establishing criteria for the equitable allocation of the rivers' waters among them and for the implementation of the principle not to cause significant harm would be an exceedingly difficult task. According to Hillel, the crucial issue is how to weigh historical rights against proportionate contributions to the rivers' flows, taking into account associated factors such as the real social and economic needs of each country; the size of each country's population; the availability of alternatives of comparable value; the efficiency of water use; and the effects of the uses of the water resources by one riparian state on the other riparian states. Hillel believes that defining criteria to deal with these factors would be formidable 'even if the contenders were willing to submit their claims to impartial adjudication', but that 'in the absence of such willingness, the issues remain in contention and may lead to armed conflict' (1994: 103).

However, as we shall see in the next section, such scepticism is unfounded. This is because, despite some provocative rhetorical statements by some officials in the region and their sensationalist coverage by Western journalists, the most important water projects on the Euphrates and Tigris rivers were carried out peacefully during the 1970s and 1980s and those which are still under construction are of far less significance. This is not to deny the weaknesses in this process – that is, separate planning, and lack of coordination among the riparian states – and its adverse impact on the level of exploitation of the Euphrates basin and on the legal positions of the states involved. However, it should be noted that the overall process of water resources development in this basin during the last few decades has been in accordance with previous agreements between the riparian countries concerning flood control and the regulation of river flow which were the main concern of the downstream states in the first phase. The only difference is that in the past the downstream states made no contribution to these highly beneficial projects.

140 Water Politics in the Middle East

Table 5.2 Per capita surface water availability in the Euphrates–Tigris basin (1990 [1995])

Riparian states	Total water/year (bcm)	Population (millions)	Per capita water (m^3)
Turkey	100	55 [65]	1818 [1538]
Syria	23	13 [15]	1769 [1533]
Iraq	91.2	17 [20]	5364 [4560]

Source: Lowi 1995; calculation for 1995 has been done by the authors

Although Turkey was motivated by its own interests, the construction of huge storage dams in that country has had a positive effect on Turkey's downstream neighbours. For instance, during the 1988–9 water year, which was the driest of the last 50 years, the deficiency in the natural flow was compensated by water from the Keban and Karakaya reservoirs protecting Syria and Iraq from the dire consequences of the drought (Kliot 1994: 137). Similarly, in the years of plenty, these reservoirs regulate the flow and protect the downstream countries from destructive floods. This is what Iraq was trying to achieve in 1946, and was ready to pay for, when it signed the Treaty of Friendship and Good Neighbourly Relations with Turkey.

Moreover, although all the water resource claims and counter-claims of the three riparian states are too complex to be resolved quickly, they have succeeded in addressing at least some of the problems through bilateral agreements. While Turkey is obliged to send no less then 16 bcm/yr of water downstream, Syria and Iraq have agreed that the Euphrates waters be allocated between them with 42 per cent to Syria and 58 per cent to Iraq (Kliot 1994: 149; Shapland 1995: 305). If cases of cooperation that fall short of formal agreements on water-sharing are included, the extent of cooperation is much greater. Furthermore, since the establishment of the Joint Technical Committee for Regional Waters in 1980, a latent structural cooperation exists among the three riparian states. This committee provides a forum for the three states to explain their different points of view, to engage in technical consultations on water issues, and to resolve any friction before it ends in conflict. In this forum, Turkey, which is in a superior geographic and military position, has tried to engage Syria and Iraq in ministerial talks on the water issue and so far has successfully contained any disputes. These measures have ensured that Syria, a riparian party with the smallest population and least economic power, is unlikely

Table 5.3 Annual surface water and withdrawal in the Euphrates–Tigris basin

	Annual river flows			Annual withdrawal		
	Internal flow (bcm)	In flow (bcm)	Out flow (bcm)	Total (bcm)	Per capita (cm)	% of total water resources
Turkey	196	7	69	15.6	317	8
Syria	7.6	27.9	30	3.34	449	9
Iraq	34	66	n.a.	42.8	4 575	43

Source: World Resources Institute (1990: 331)

to let frustrations over water get out of hand. This suggests that downstream states in the Euphrates–Tigris basin would not resort to military force to prevent threatening developments upstream if water were the only point of disagreement.

The last point which should be noted here is that of data anomaly (see Table 5.2, and 5.3). The preceding section showed that there are wide discrepancies between different accounts of the hydrological data concerning the Euphrates–Tigris basin. This is also the case with regard to the current levels of water extraction for irrigation and plans for development neither of which are known with any precision. Such data anomalies in the available records concerning water and irrigable land resources in the Euphrates–Tigris basin can mislead observers and undermine their conclusions (Bilen 1994: 110–12; Kliot 1994: 133–50, Kolars 1994: 75, 81, 85). The discrepancy between the total water resources of each country in Tables 5.2 and 5.3 is an example of the kind of data anomaly which is noticeable in many reports and studies. In both tables, Iraq seems to enjoy the highest rates of per capita water availability and withdrawal, but this is partly because of erroneous calculations such as double entry accounting, and partly because those who compiled the data are not familiar with the real situation in the region. The figure concerning per capita water withdrawal in Iraq is seriously misleading because the huge amount of water which drains off to the Gulf from the Euphrates and Tigris Rivers, through Shat al-Arab, is not accounted for.

Underlining the significance of the acquisition, verification and analysis of data for a rational water management system, Kolars (1994: 88) declares that, 'Data regarding stream flow, precipitation, evapotranspiration, water removals, return flow, salinity and a host of other variables are notoriously scarce, incomplete and open to question everywhere in the Middle East.' He believes the main reason for scarcity or anomaly of hydrological data in this region is that, 'Nations have, until now, viewed data as knowledge, and by extension, data as power.' Drawing a parallel between obtaining hydrological data and counting sheep in the desert, Kolars contends that, "If a fodder supplement is being offered there are plenty of sheep. If a head tax is proposed, there are very few sheep" (1994: 88). This is why, despite bilateral agreements between the three riparian states to monitor the system and exchange data on river levels and discharges, secrecy, obscurity, and confusion surround hydrological data in the region.

Fortunately, however, the riparian governments have understood that successful negotiations for a long-lasting water-sharing agreement depend on accurate hydrological data. In this regard, remote sensing technology is of great importance; several countries in the region are now trying to obtain data acquired through satellite photography. While the Arab countries could benefit from the data acquired by the Geographical Information System (GIS) commissioned by the government of Qatar, Turkey has been negotiating with the Earth Observation Satellite Company (EOSAT) to provide it with an archive of past satellite imagery which contains recent historical coverage of natural conditions in Turkey, the adjacent Middle East, and the Central Asian Republics (Kolars 1994: 89). Technically, it is not difficult to coordinate these efforts, but politically it seems formidable, as Kolars observes.

Will the riparian countries of the Euphrates–Tigris basin fulfil such a demanding task as establishing a comprehensive international water agreement through negotiation, cooperation, and compromise, as the main hypothesis of this study suggests? Or are they following a course of conflict-orientated polices that will end in more secrecy, complexity, confusion, conflict, and war, as is conventionally presumed? The next section examines the latest developments in the riparian countries to see what is the prospect for hydropolitics in this basin. Since from a hydrological point of view it is Turkey, master of the vast majority of the rivers' flow, whose hand is on the tap, we begin our analysis from the upstream developments and conclude with the downstream developments. What is visible in this journey is that all three countries are undergoing rapid change, and their considerable expenditure on major water projects indicates that all appreciate water's future crucial role.

Turkey's water policy and Arab reaction

Conception of water as a strategic issue

In any situation in which a catchment is divided between a number of states, upstream abstraction will affect the quantity and quality of water available to downstream users in one way or another and there is always the possibility of friction. Even if there is some agreement about sharing, control always remains in the hands of the upstream states(s), and thus politically motivated actions can never be discounted. In a volatile political situation, the absence of a clear-cut treaty regulating the methods of sharing the waters between the riparian parties encourages suspicion, confusion, and misperception. In such circumstances any significant increase in the ability of upstream countries to control the headwaters could easily evoke annoyance and tension in downstream countries. Appropriate institutions could help the parties to rectify their misperception and alleviate tensions and anxieties peacefully, but this ideal solution is not the only remedial measure; if the parties have efficient leverages to persuade each other that they will be better off with cooperation, they could mutually prevent their frustrations from getting out of hand. This is exactly the case in the Euphrates–Tigris basin.

As mentioned above, the water-related agreements which were concluded during the mandatory period and in subsequent years between the riparian parties indicate that the parties accepted the principle that hydrological schemes for the benefit of downstream states may be constructed in the upstream state. This principle was particularly manifested in the 1946 Treaty of Friendship and Good Neighbourliness between Iraq and Turkey. Such a spirit of cooperation prevailed until the 1960s, when radical nationalist forces created a ferment that led to frequent military coups and finally ended up in the seizure of power by rival factions of the Ba´ath party in both Syria and Iraq. Advocating a policy of socialism and Arab nationalism and a foreign policy orientated towards the USSR, the ruling elites of this faction-riven party began to watch each other and Turkey with suspicion. Turkey, as a member of NATO and the host of US military bases, was no longer perceived as a reliable partner and, therefore, upstream water development schemes were regarded with suspicion by downstream parties because they might give too much control to Turkey. As a result, from this time onwards, Syria and Iraq declined to give their consent to Turkish water projects on the Euphrates and Tigris. By the same token, Iraq was concerned about Syria's water development plans and this issue sometimes stirred up tensions between the two countries.

All three countries had ambitious plans to develop their water resources and expand their hydropower and agricultural outputs. Turkey's decision to use the water resources of the Euphrates–Tigris basin was reinforced in the 1960s when the need for electric energy emerged as the most urgent national priority. Preliminary work on the Keban Dam, with a storage capacity of 30 bcm, began in 1964–65.[11] This was the first major Turkish project on the Euphrates and it was followed by a massive water management scheme that involved dam building, diversions and the extension of irrigated agriculture in the southeastern part of the country, known by its Turkish acronym GAP (for *Güneydogu Anadolu Projesi*).

According to the Turkish official records (GAP Web Page 1997)[12], the GAP is an integrated multisectoral master plan which includes 13 major projects and envisages the construction of 22 dams and 19 hydroelectric power plants on the Euphrates and Tigris rivers and their tributaries. It is planned that, at full development, over 1.7 million hectares of land will be irrigated and 27 billion kWh of electricity will be generated annually with an installed capacity of over 7500 megawatts. The area to be irrigated accounts for 19 per cent of the economically irrigable area in Turkey (8.5 million ha), and the annual electricity generation accounts for 22 per cent of the country's economically viable hydropower potential (118 billion kWh).[13] Before the construction of the Keban Dam, Turkey used only 3 per cent of the Euphrates water, and the first two dams of the GAP caused only minor water loss from the river. In fact, these two dams simply regulated the fluctuations of Euphrates discharge (Kliot 1994: 136). However, the total water withdrawal for irrigation associated with the Euphrates portion of the GAP was projected in the master plan to amount to 10.429 bcm/yr. This rang alarm bells in both Syria and Iraq who rely heavily on the Euphrates River for drinking water, irrigation, industrial uses, and hydroelectricity, and view any upstream development with concern. Their anxiety increased by exaggerated predictions suggesting that Turkish water extraction would exceed 16.9 bcm/yr (Kolars and Mitchell 1991: 208 and 212; Kolars 1994: 74–5).

The impact of the GAP on Syria and Iraq has been extensively covered in the literature (for an instructive study see Kolars and Mitchell 1991). It was clearly to the advantage of Syria and Iraq to reach agreements with Turkey that would guarantee them a fixed quantity of water each year, or a fixed percentage of available flow. However, adopting the legal doctrine of *absolute sovereignty* (see Chapter 2), Turkey was reluctant to give its resentful neighbours an established easement which would affect its most important national development plan – the GAP. Neither was Turkey ready to give Syria and Iraq a role in the GAP, nor were they likely to respond to Turkey's calls for regional coordination without an explicit agreement ensuring that the management and sharing of rivers was equally in the hands of all the riparians involved. Since Turkey, consequently, had decided to pursue its own interests, Iraq and Syria took two types of precautionary measures to safeguard their access to water resources. First, they embarked on their own respective national plans with the aim of increasing their strategic storage capacity by impounding the rivers' water in gigantic reservoirs. Second, they initiated an international diplomatic campaign to prevent, modify, or delay the unilateral upstream projects as much as possible.

A total storage capacity of 15 bcm was planned by Syria – about 12 bcm of which has been provided by the Tabqa (al'Thawra) Dam. Iraq, so far, has accumulated a storage capacity of 100 bcm, including barrages. With her three major dams on the Euphrates – Keban, Karakaya and Ataturk – Turkey's current storage capacity is estimated at 90–100 bcm (Hillel 1994: 103; Kliot 1994: 122, 137–8). Such a tremendous build-up of storage capacity, which is far beyond the actual needs of the three riparian states, has been widely tipped as a key factor that will intensify competition and conflict among the parties. The argument is that, 'large storage capacity tends to persuade co-riparians to prefer the accumulation of waters within their own territory instead of sharing it with their partners' (Kliot 1994: 122), and since this capacity greatly exceeds the annual flows of the two rivers, there is always a perceived shortage of water and, by extension, a constant risk of conflict. Accordingly, Starr (1991: 31) claims, 'Unilateral construction of new dams could lead to escalating disputes and armed confrontation.'

However, this is not necessarily the case in the Euphrates–Tigris basin where, for both Turkey and Syria, the imperative of hydropower is as strong as the needs of agriculture. Since the two Turkish giant reservoirs of Keban and Karakaya are designed for hydropower production, they operate as a water regulator rather than a storage bank and they are very beneficial to the downstream countries.

The third Turkish reservoir, which is the most controversial and the masterpiece of the GAP – the Ataturk Dam – and the Syrian reservoir of the Tabqa Dam, are dual purpose and at these points there is the highest risk of water extraction for irrigation. However, the two countries will inevitably find themselves enmeshed in a nexus of competing national demands for water: the more agriculture, the less power.[14] Taking account of their increasing energy requirements, both Turkey and Syria must decide between agriculture and electricity. Less irrigated agriculture means that more water could be used for power generation which in turn would offset the need for petroleum and save much-needed hard currency. By contrast, more irrigation entails less water for power generation. Moreover, more irrigation requires more electrical energy to pump water to additional, higher fields, and this adds to the costs of agriculture and increases the need for energy imports. Thus, as Kolars (1994: 72–3) rightly noted, at stake are the questions of foreign exchange, petroleum imports, provision of inexpensive energy for industry, and the assurance of sufficient and economical food for home or sale abroad. Moreover, the international aspect of the issue is of considerable importance. Since hydropower production is a *nonconsumptive* water activity, more power generation minimises the political and environmental impact on downstream countries. But if irrigated agriculture, which is excessively water *consumptive*, is prioritised, then downstream agitation will increase and lead to retaliatory action which would adversely affect the national interests of the upstream country.

The construction of enormous water storage capacity in Syria and Iraq has partially enhanced their sense of water security. However, it was their diplomatic conduct that progressively encouraged Turkey to adopt a conciliatory and cooperative attitude. Indeed, Syria and Iraq took every opportunity offered by diplomacy to prevent upstream developments or at least to modify them. They challenged Turkey over the water issue from all aspects, using their legal, financial, and security leverages. Thenceforth, water diplomacy became part and parcel of the regional foreign policy of the three states. In the following section we shall discuss the impact of the water issue on their respective foreign policies.

Water diplomacy

Given the lack of trust and compatibility between the three riparian states, the idea of developing the Euphrates–Tigris basin as a unified whole for the greater good of all parties seemed elusive. The downstream states sought an explicit agreement which would guarantee them a secure share of water each year, but despite the agreements of the mandatory period and the Treaty of Friendship and Good Neighbourly Relations that established general principles pertaining to the right of downstream countries to waters entering their territory, Turkey refused to discuss the issue of water division, arguing that it had no legal obligation to do so (Frankel 1992: 7). The Turks claimed that since there was no international regulation which governs the rights of riparian states, nor any specific treaty which regulates the sharing or common exploitation of the Euphrates and Tigris, the status quo was in Turkey's favour. However, Iraq and Syria insisted on their right to share the waters of the rivers. Their major argument was that international precedent, if not enforced law, warrants that the management and sharing of international rivers be equally in the hands of all the riparian countries involved (Kolars 1994: 88). Referring to their ancient 'acquired rights', because they made the 'prior use' of the water resources, Iraq claimed 59 per cent of the natural flow of the Euphrates at the Syrian–Iraqi border and Syria claimed 40 per cent of the flow at the Turkish–Syrian border (Bakour and Kolars 1994: 139; Wakil 1993: 64).

Turkey's response was fourfold: first, adopting the legal doctrine of *absolute territorial sovereignty*, Turkish sources argued that the Euphrates and Tigris both originate on Turkish soil and are Turkish rivers while they flow over Turkish territory, concluding that Turkey is not obliged to share its waters with its Arab neighbours. Second, Turkey claimed the Euphrates and Tigris as 'transboundary' rivers whereas Syria and Iraq considered them to be 'international'. Third, it maintained that the Euphrates and Tigris rivers must be considered together as a single transboundary water course system. Fourth, Turkish officials declared that Turkey would agree to share transboundary waters *if* they included the river Orontes, which flows through Lebanon, Syria, and Southern Turkey to the Mediterranean Sea, as well as the Tigris river. As we shall see, this Turkish position plunged them into an abortive legal controversy at regional and international level.

In short, Turkey claimed an undisputed right to use the water within its territory without the consent of other riparians, in the same way that Arab states regard oil as a natural resource which is subject only to their sovereignty. This Turkish attitude was clearly set out at a news conference by Suleyman Demirel, then prime minister, saying that:

Neither Syria nor Iraq can lay claim to Turkey's rivers any more than Ankara could claim their oil. This is a matter of sovereignty. We have a right to do anything we like. The water resources are Turkey's, the oil resources are theirs. We don't say we share their oil resources, and they cannot say they share our water resources. (Bulloch and Darwish 1993: 74–5)

This controversial argument brought up the question of the legal status of water and the legitimacy of drawing a parallel between the legal status of oil and water (see Frankel 1992: 7; Chalabi and Majzoub 1995: 208). There was a firm response from Arab countries. For instance, Egyptian sources declared that the Turkish position was contrary to international law and justice and incompatible with good neighbourliness and friendly relations, and argued that it is illegal to deny the peoples of the neighbouring countries the water resources they have been using for centuries. Syria adhered to the doctrine of *limited territorial sovereignty* and called for a fair sharing of the Euphrates and Tigris waters, saying that, 'Nobody has the right to divert these rivers and subject Syrians to a catastrophe. If every country started to divert rivers claiming they were on their lands then the whole world would be subject to grave dangers' (Bulloch and Darwish 1993: 75). For its part, Iraq held to the doctrine of *absolute territorial integrity*, insisting on its ancient or prior rights of usage of water from the Euphrates and Tigris rivers (Kliot 1994: 123). These sharply divergent points of view were actively prosecuted by all three parties in international arenas especially in the process of codification of international law which led to the conclusion of the Convention on the Law of the Non-navigational Uses of International Watercourses. During this process Turkish officials repeatedly expressed reservations as to the creation of an international regulatory framework for cross-border rivers (Chalabi and Majzoub 1995: 210).

When Turkey realised that its adherence to the doctrine of *absolute territorial sovereignty* was no longer acceptable to the body of international law, which treats the river basin as an undivided unit, it suggested that a clear distinction should be made between the concepts of 'international' river and 'transboundary' river. Turkey defined the 'international river' as a river which has its opposing banks under the sovereignty of different countries, whereas the 'transboundary river' is a river which crosses common political borders.[15] Accordingly, the Euphrates and Tigris must be considered as transboundary rivers. It then argued that waters of international rivers must be shared by the riparian countries through the median line while waters of transboundary rivers have to be utilised in an equitable, reasonable and optimal manner (Tekeli 1990: 212–13; Chalabi and Majzoub 1995: 211). However, this Turkish classification of the Euphrates and Tigris was acceptable neither to its downstream neighbours nor to the main body of the international community. It was clearly at variance with the definition of the international river basins contained in Article 2 (b) of the UN Convention according to which 'International watercourse means a watercourse parts of which are situated in different States'.[16] Once this is established, the problem regarding the Euphrates and Tigris turns to a controversy over the scope of the rights of each riparian state to the part of the basin within its territory.

Turkey also asserted that the waters of the Euphrates and Tigris form a single transboundary water basin, thus allowing Tigris waters to be used as a substitute for Euphrates waters. This argument is based on the facts that both rivers originate from a similar geographical area, scarcely 30 km from each other in Turkey, they are linked by artificial canals such as the Tharthar Canal in Iraq; and they naturally join each other further downstream to form the 120-mile-long Shat al-Arab water way. However, Iraq claims that the Euphrates and Tigris are two separate entities and hence should be treated as two distinct river basins (Kliot 1994: 163; Chalabi and Majzoub 1995: 211). Insisting on these criteria, each time that the Iraqi and Syrian representatives express the importance of concluding a treaty concerning the sharing of the waters, the Turkish officials reply, 'agreed, on condition that the sharing of transboundary waters includes the river Asi (Orontes)[17] which successively flows across Syria and Turkey, as well as the Tigris' (Chalabi and Majzoub 1995: 228). While Iraq resents the idea of treating the Tigris and Euphrates as a unified system, Turkey's reference to the Orontes brings in the thorny issue of its historical territorial dispute with Syria over Hatay province (see preceding section). If a general water agreement were to cover the Orontes, both the Syrians and the Turks think that it would imply the recognition of Hatay as Turkish. While Syria has regularly referred to this disputed territory as the 'Arab Iskenderun' and thus insists on the *Arab* character of its waters (Ibid.: 229), the Turks argue that:

If comparison is made between the utilisation of the Orontes and the Euphrates, there is justified cause for Turkey to complain about how the water of the Orontes is completely consumed by Syria and Lebanon, while Turkey releases 500 cm/s of water even when the velocity of the Euphrates falls to 100 cm/s. (Bulloch and Darwish 1993: 69)

Regarding the controversy over the legal status of water during the codification of the UN Convention on the Law of the Non-navigational Uses of International Watercourses, all parties gradually recognised that the two extreme points of view, *absolute territorial sovereignty* and *absolute territorial integrity* must be moderated by the concept of 'equitable and reasonable utilisation' of transboundary rivers, with a view to attaining the optimal and sustainable utilisation of those waters, recognising both the duty to cooperate in the protection of the watercourse system, and the obligation of riparian countries 'not to cause significant harm' to each other (Articles 5 and 7 of the Convention). Turkey understood that the concept of absolute sovereignty is now obsolete even for those parts of an international basin which lie exclusively within the territory of one riparian country. Turkey also acknowledged that qualified sovereign rights have gained general acceptance together with the notions of interdependence and international harmony, and are emerging as fundamental bases in the behaviour of the modern community of nations. The substitution of unrestricted rights by qualified rights means that the attributes of sovereignty can no longer justify arbitrary usage and, in the absence of an established historic right, the common interests of all riparian parties form the basis of their legal rights (see El Morr 1995: 296). On the other hand, Iraq and Syria discovered that their being downstream states which were first to develop a shared water body was not sufficient to foreclose subsequent upstream developments by signifying that the latter developments would cause them harm. Under the doctrine of equitable utilisation, the fact that a downstream state was 'first to develop', and thus had made prior uses that would be adversely affected by new upstream uses, would be merely one of the factors to be taken into consideration in arriving at an equitable allocation of the uses and benefits of a shared watercourse (McCaffrey 1992: 17–29; Bilen 1994: 114).

The convention emphasises a balance between the interests of the states concerned, with each state making use of the water in an 'equitable and reasonable manner', while avoiding causing 'significant harm' to other riparians – a principle very much favoured by downstream states. To determine what is 'an equitable and reasonable share', the convention suggests that all relevant factors must be taken into account, including the physical and climatic conditions of the entire river basin; the comparative economic and social gains enjoyed by each state and by the entire river community; the dependency of the population of each riparian state upon the waters; the extent to which the needs of a riparian can be met without substantial damage to others; prior uses of water resources; economy of use; and the availability of alternatives (Article 6.1).[18] Although Article 7 requires all riparian states to take all appropriate measures to prevent the causing of significant harm to each other, there is no explicit definition of what harm would be significant, and what would not. However, the second paragraph of the same article advises that, if significant harm is caused to a riparian state, the state(s) whose use causes such harm shall, in the absence of agreement to such use, take all appropriate measures to eliminate or mitigate such harm and, where appropriate, to discuss the question of compensation.

These considerations illustrate how the rules of public international law place all the riparian states on a basis of mutuality and equitable participation. The convention, as a comprehensive global code of law in this field, establishes a regime of reciprocal obligations and fair apportionment of the international water resources based on the duty of the riparian countries to contribute on an equitable basis to the protection and control of the system as particular conditions warrant or require (Article 8). As such, the convention seems to have some restraining effect on upstream states and supports the downstream countries to present their case against unilateral upstream developments. However, the principle of optimal and sustainable utilisation of the water resources and the duty of all parties to participate in the development and protection of the water basin provides a ground for upstream states to question the wasteful and uneconomic uses of water in downstream states and ask for their contribution to the cost of upstream projects which are beneficial to all parties and to the protection of the ecosystem. It is here that Turkey has recently attempted to find a real solution to the question of water and declared herself ready to cooperate.

At the same time as claiming that it has abided by the UN-endorsed principle of equitable, reasonable and optimal utilisation of the transboundary water resources, and caused 'no significant harm to other riparian parties', Turkey has accused its downstream neighbours of wasting huge amounts of water through their primitive agricultural techniques or by using water to irrigate non- or low-productive lands (see Tekeli 1990: 214; Frankel 1991). Moreover, Turkey claims it has single-handedly undertaken costly water projects which have greatly benefited both downstream states and protected the entire ecosystem. According to Kliot (1994: 163), during negotiations with Syria and Iraq who demanded a constant Euphrates flow of 700 cm/s, Turkish officials argued that such a demand was for a regulated release of Euphrates flow without any Syrian–Iraqi contribution to the expensive facilities required for such a regulation. Since 500 cm/s of controlled water is much more valuable to downstream countries than a much greater quantity of uncontrolled water, as Kliot (1994: 164) points out, Syrian and/or Iraqi compensation for Turkey's investments in dam construction for the upper Euphrates is well within the accepted rules of international law and could persuade Turkey to release larger amounts of water. This argument is clearly evident in the remarks of several Turkish politicians and specialists gathered in a comprehensive set of interviews conducted in December 1990 and January 1991 by Norman Frankel (1991), who examined Turkish perspectives on water management and the influence of water on Turkish domestic and foreign affairs. It is noteworthy that, although the interviewees have different political orientations, none of them suggests that water is going to be a source of conflict between Turkey and its neighbours. On the contrary, they all see the present indignation of their downstream co-riparians as a passing phase and emphasise the feasibility of co-operative arrangements between riparian states for establishing a regional water management system.

Syria and Iraq, along with their legal campaign, initiated a financial campaign against the upstream Turkish projects. Securing an international financial blockade on the GAP, they successfully forced Turkey to bear the huge cost of its major projects single-handedly, a process which adversely affected the Turkish national economy, and made it postpone the full implementation of its projects for at least three decades.[19] Since the GAP was too costly to be financed by the government, Turkey needed to seek funds from international lenders or donors. However, the World Bank, the most important international lender in this field, has a firm policy concerning projects on international water basins (World Bank 1993; Le Moigne 1994b;

Goldberg 1995; Krishna 1995). Basically, the Bank does not approve an application for funding a project if another riparian objects. However, although this policy plays directly into the hands of the downstream parties, it does not give them an absolute right of veto. The Bank examines the basis of the objection and the possible effects of the project on objecting riparian parties and only if it finds the objections to be substantial will it decline to fund the project. The irony is that it was the Turkish objection to Syrian projects on the Orontes (Asi) river in the 1950s that led the World Bank to develop an operational policy on international river basins (Krishna 1995: 34). Other international lending institutions, such as the African Development Bank (ADB) and the European Investment Bank (EIB), work on similar if less explicit guidelines. The governments of major donor countries are more flexible than international institutions. However, in the absence of political preferences, they also would prefer not to help one country at the expense of another.

When Syria protested against the Karakaya Dam project in Turkey, the Bank dismissed its objections on the ground that the dam was a hydro-power project which would have regulated but not reduced the river's flow (Shapland 1995: 310). But Syria and Iraq won backing from the Arab League, the oil-rich Arab countries of the Gulf region, and other international lending institutions not to finance the Turkish GAP project. They also urged foreign contractors not to cooperate with Turkey on this project, and threatened them that otherwise they would be shut out of future projects in Iraq and Syria and perhaps in other Arab countries. Faced with this pressure, the World Bank announced that it would support the GAP only if it was formally accepted by the other two riparian states. As a result, since Syria and Iraq refused to accept it, apart from external export credits towards electricity generating equipment, Turkey has had to bear the huge cost of the GAP from its own hard-pressed budget. This partly explains the numerous delays and scale-backs of the project.

Moreover, as Bulloch and Darwish (1993: 65) point out, every lira had to be funded by Turkey at a time when inflation was running at 70 per cent and unemployment was rising. Since Turkey was unable to employ foreign contractors, the Turkish policy-makers portrayed the GAP as a matter of national pride, encouraging Turkish companies to prove themselves capable of completing the project. But the project distorted the national economy of Turkey, strained government finances, fuelled inflation of more than 100 per cent, and led to a severe financial crisis in 1994. The cost of the GAP has now reached 32 billion dollars and is expected to add more than one billion dollars a year to Turkey's GDP (*Financial Times*, 10 November 1994: 8). Many Turkish economists blamed the GAP for many of the country's economic troubles, and believe that the priority given to this project was the result of an obsessive interest in it by some politicians, such as Turgut Ozal and Suleyman Demirel.

However, although the financial leverage exerted by Syria and Iraq significantly decreased the pace and intensity of development in the GAP region, it did not deter Turkey from financing the project, particularly the Ataturk Dam which is its centrepiece, from its own sources regardless of its downstream implications. Turkey's failure to appreciate the gravity of the water issue to its downstream neighbours tempted Syria and, to a lesser extent, Iraq to take a more active interest in the internal affairs of their upstream neighbour as a way of reminding Turkey that it would be better off with cooperation – and the thorny issue of Kurdish insurgency was the most appropriate internal Turkish problem.

The Euphrates and Tigris rivers rise in Kurdish territories where people despise the government because for many years it has not only fiercely suppresses their long-held desire for autonomy, but has also denied their identity and systematically set out to eliminate their culture and language (Gunter 1990). Syria realised it has a potentially strong card to play – that of security. The national security and the stability of Turkey were vulnerable to Kurdish sedition and secession. So Syria decided to use this security leverage to encourage Turkey to take account of the demands of its downstream neighbours. For Syria, the water issue was not only an issue in itself; it was also a bargaining chip in negotiations with Turkey about their territorial dispute over Hatay province. Iraq was not as determined as Syria for three reasons: first, Iraq itself was exposed to a similar risk, since it had its own Kurdish problem; second, during the 1980s it was engaged in an all-out war with Iran and the water issue with Turkey was not one of its highest priorities; third, Iraq desperately needed to have a good relationship with Turkey, through whose territory its oil could reach the market.

Syria's decision to step up support for the Kurdish guerrillas of the Kurdistan Workers' Party (PKK) ignited a rebellion that is now by far Turkey's biggest domestic security problem, tying up the Turkish army in the region and imposing much more pressure on the already drained national budget (Bulloch 1993: 12; Barham 1994: 8). At first, Turkish officials maintained that water rights were not to be negotiated as a trade-off against the PKK. However, they gradually realised that they had to take into account what their neighbours could do to affect the situation inside Turkey by giving or withholding support for the Kurdish rebels. The GAP, a project which was supposed to bring prosperity and stability to a backward rebellious region, was turning into a destabilising factor and ruining the Turkish economy's chances of stable development – the economy slumped in deep crises and Turkey's big cities were overwhelmed by waves of migrants fleeing the fighting in the Kurdish southeast (Barham 1995: 2). Turkey acknowledged that the Kurdish issue, which is intrinsic to Turkish national security and stability, had become subordinate to the water issue since this issue was at the heart of Turkey's relations with Syria. A senior official of the Turkish foreign ministry was quoted:

> It is true that Syria does have a habit of working through proxies. It was about 1980 that we started talking very seriously about expanding the GAP project, and it was about that time that Ocalan [the leader of PKK] began getting help from Damascus. You could make a connection. (Bulloch and Darwish 1993: 71)

It was now clear that Syria would support the PKK if it thought Turkey was inflexible over the water issue. The acknowledgement of this fact led to negotiations that had as their bottom line the exchange of water for security. In 1986, when the Syrian prime minister paid an official visit to Ankara, for the first time the Syrians publicly linked security and water saying that 'they would sign the security protocol only if Turkey entered into a formal water agreement' (Bulloch and Darwish 1993: 66). In June 1987, the Turkish prime minister, Turgut Ozal, visited Damascus and signed the security protocol and promised to maintain the Euphrates flow of 500 cm/s at the Syrian border – this was Turkey's second retreat from its initial position on the Euphrates since 1964.[20] This commitment was repeatedly confirmed by Turkish authorities and finally, in April 1994, an agreement was signed by their heads of state, in which Turkey agreed to supply Syria's water requirements in return for Syrian assurance of cooperation on the Kurdish problem (Starr 1995: 137).

Many sceptics have questioned the credibility of this agreement, arguing that it has been written in 'invisible ink' and 'it was not the permanent agreement Syria was hoping for' (Starr 1995: 137).

However, there are indications that, despite fanciful predictions of the exacerbation of water conflict in this basin, Turkey is reshaping its water policy by trying to find a conciliatory solution to the region's water problems. This is because Turkish policy-makers have realised that they will be better off with cooperation, however difficult that may be. To show its goodwill, Turkey has not only declared that it will renounce permanently half the flow of the Euphrates, but it has also proposed to bring drinking water from Anatolia to the Levant and down to the Arabian Peninsula through a pipeline to help the water poor countries in this area (see Chapter 6). As Kolars (1994: 78) rightly acknowledges, Turkey's proposal of a 'Peace Pipeline' was an important cooperative gesture which may become part of the milieu in which future river management will be forged, though there are question marks over its economic or political feasibility. This proposal also underlines the fact that Turkey is a Middle Eastern country with a water surplus, and therefore is capable of smoothing regional inequities by providing water for international sharing.

This recent Turkish concession nullifies the assumption that the relationship between upstream and downstream states is purely one way and that, given the superior position of upstream states, the interests of downstream countries can be ignored. This assumption is the basis of many arguments for 'water conflict' in the international river basins (see for instance, Falkenmark 1986a; Starr 1991; Lowi 1995). It also refutes the hypothesis that 'whoever controls water or its distribution can dominate the Middle East and all its riches' (Bulloch and Darwish 1993: 161). Moreover, the current mood of water management policies in the region indicates that a new paradigm is emerging in which not only have qualified sovereign rights gained general acceptance together with the notions of interdependence and international communion, but also previously ignored environmental issues are beginning to be taken [seriously] into account (see below).

In the process of their diplomatic exchange, all parties have understood that the international water resources cannot be monopolised or subjected to the control of a single riparian state. This active, continuous, and critical interchange has also convinced them that cooperation over the water issue is not necessarily a zero-sum game but could be a win–win situation in which all parties would gain from cooperation. Turkey, whose hand is on the tap, needs a water-sharing agreement with its riparian neighbours in order to secure financing from the World Bank and other international lenders and donors to complete the GAP. It also needs to acknowledge the equitable share of its downstream neighbours whose annoyance can affect the situation inside Turkey and undermine its socioeconomic development, security, and stability. The benefits to Syria and Iraq of an agreement that delivers water security are obvious.

With the end of the Cold War, the obsessive preoccupation of states with a competitive notion of military security and political adversarialism is giving way to a common realisation of the need for cooperative environmental security. All three riparian states of the Euphrates–Tigris basin have realised that politically-motivated development plans are self-destructive and should be modified in accordance with economic and environmental criteria. In Turkey, questions regarding the economic and environmental impact of the GAP are raised by the newly established Environmental Issues Foundation (*Turkiye Cevre Sorunlari Vakfi* – *TCSV*) (Kolars and Mitchell 1991: 76). Among the issues addressed by this group are the considerable amount of Turkey's most fertile lands which will be inundated by the new reservoirs; the environmental side-effects of increased irrigation and fertilisation; the significant increase in demand for equipment by mechanised agriculture; the role of market conditions in determining what crops should be cultivated and a host of other questions which will bring a vast spectrum of social and economic challenges to the GAP region and, by extension, to Turkey as a whole. These considerations will cast a shadow on the future of the GAP in Turkey.

In Syria, despite initial estimates that the Tabqa (al´Thawra) Dam would increase the irrigated area to 600 000–650 000 hectares (Naff and Matson 1984: 91), the revisions in Syria's agricultural plans have reduced the projected area to 240 000 hectares (Kolars 1994: 50) and the government's attention has turned to its rain-fed lands, which still account for 80 per cent of the country's cropped area (Starr 1995: 132). The political imperative of the Cold War era did not permit the Syrians to do an accurate soil survey in the vicinity of the dam site and, therefore, they failed to predict the catastrophic effect of gypsiferous soils both on canals and on field applications of water. Now that the Syrian policy-makers have recognised that the cost of land reclamation for the project is so high, they place greater emphasis on dry farming and rain-fed agriculture in the narrow coastal belt which benefits from abundant rain.

As for Iraq, it has now admitted that improving its current irrigation networks, increasing irrigation efficiency through the application of better technology and water conservation methods, and transferring water from the Tigris to the Euphrates are the most practicable options to mitigate any impending water crisis (Hardan 1993: 79). Before the Kuwait Crisis, Iraq was planning to invest over $300 m. in more than 20 water projects on the Tigris river and the Tharthar Canal – a major reservoir which links the two rivers and enables Iraq to compensate herself for any contingent water deficit of the Euphrates (see Map 5.1). This is what Turkey and Syria would like Iraq to do. However, another Iraqi project which was designed to drain the agricultural fields of Southern Mesopotamia and rehabilitate 1.5 million hectares of barren land over five to ten years was more controversial. In December 1992, the Iraqis celebrated the inauguration of a 565 km-long navigable canal which flows midway between the Euphrates and Tigris – called the 'Third River' (see Map 5.2). It starts near Baghdad and ends close to Basra in the south and is designed to reclaim polluted land by removing 80 tonnes of salt a year. Although this project was initially planned and supported by Western aid, after the Persian Gulf War a suspicion arose that the Iraqi regime intended to destroy the entire ecosystem of the marshlands which had long been a buffer zone for Iraqi dissidents, including the Marsh Arabs. Such suspicion led the media and UN officials to call the canal 'an environmental crime'. However, the Iraqis deny the charge that they have any intention of destroying the natural ecosystem of the marshlands.[21]

Conclusion

The discussions in this chapter clearly indicate that in contrast to what is conventionally believed, first, water resources have never been the root cause of military conflict in Mesopotamia and, second, since antiquity, hydraulic civilisations which flourished in the Euphrates–Tigris basin have been forced to cooperate and coordinate their collective efforts in a systematic way in order to control the two mighty rivers for the sake of all beneficiaries. This argument is supported by several hydrological and historical facts.

First, the annual discharge of the two rivers has been more than enough to provide for the needs of all riparian communities. Although the extreme seasonal and annual fluctuations of the twin rivers have been always a great challenge to the economic development

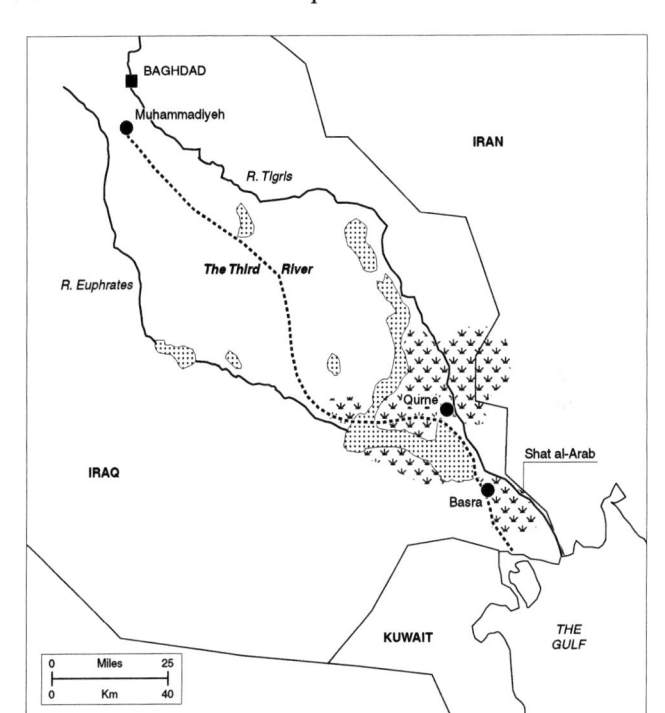

Map 5.2 The 'Third River' in Iraq

and social stability of those civilisations, it has led them towards collective cooperation and coordination to control the abundant water resources of the two mighty rivers.

Second, according to archaeological evidence, the hydraulic civilisations of Mesopotamia not only invented the most suitable tools for efficient water utilisation such as the wheel, windmill, and pipe, but also developed a remarkable water management system, through extensive networks of dikes, canals and reservoirs, which helped them successfully cope with the natural conditions of their environment.

Third, these civilisations had the social prowess and well-established legal institutions required for maintaining the functionality of their organised water systems and preventing conflict. The Code of Hammurabi in eighteenth century BCE in Mesopotamia is one of the greatest of the ancient codes and includes rules for maintaining an efficient water management system, while in the medieval era, the Islamic law evolved an extensive and elaborate set of principles to regulate water management and to abate conflict. This rich and refined literature is almost always neglected by Western observers who comment on the water resources management in the Middle East. Despite recent alarmist warnings by commentators and their conflictual representation of hydropolitics in the Euphrates–Tigris basin, this chapter shows that, in marked contrast to the Jordan River basin, none of the riparian countries is facing an imminent water shortage. Moreover, we point out that, notwithstanding the extensive water development projects of all riparian countries and the transposition of the East–West rivalry to the area which grafted itself on to the political resentments and hostilities which already existed in the region, there have been considerable efforts made by the riparian countries to avoid water conflict. Contrary to the situation in the Jordan basin, there has been no military conflict between the three riparian states of the Euphrates–Tigris basin and no violent water conflict has marked their relationship. Indeed, the three parties have been engaged in a continuous, active, and critical dialogue and technical consultations since the early 1960s. This progressive process has made the riparian states more creative, and by opening up the possibilities for more understanding, it has contributed to a higher state of cooperation among them. All this is reflected in bilateral agreements between Turkey and Syria and between Syria and Iraq whereby Turkey is committed to send no less than an average of 500 cm/s of water downstream, and Syria and Iraq have agreed that Syria receive 42 per cent and Iraq 58 per cent of the water available in the Euphrates (Bakour and Kolars 1994: 139; Kliot 1994: 149; Shapland 1995: 305). Except for a few brief periods – such as spring 1975 and January 1990 – in which the Euphrates' flow fell below 500 cm/s, usually the flow has been well above the agreed average. For instance, according to Turkish officials, in August, September and October 1994 – the driest season in the region – the flow was about 1000 cm/s (Bodgener 1994: 22).

Our analysis of water diplomacy in Mesopotamia indicates that there are several factors which strongly mitigate against the outbreak of conflict in the future. First, the actual water demand of all three riparian countries in the foreseeable future will be less than originally projected. This is mainly because of the difficulties they have encountered in meeting their land reclamation targets, and also because, more than they need irrigation water, both upstream and midstream countries need the Euphrates' waters for hydropower as a cheap and non-depleting source of energy. Second, the desire to solve the problems of waterlogging and saline deposit associated with traditional irrigation methods which is a common problem in the region, particularly in southern Mesopotamia, will encourage the adoption of more efficient patterns of water utilisation and new water-saving irrigation techniques and technologies which not only reduce estimated water requirements of all three countries significantly, but also increase crop yield. Third, the ability of Iraq to transfer the Tigris water to relieve any contingent shortage in the Euphrates is a comforting alternative. Fourth, consultations are continuing among the riparian states in the Joint Technical Committee – a body that, according to Kliot (1994: 162), reflects a cooperative trend among the three riparian states and has evolved into an active organisation which deals with all the water issues among the countries. Last but not least, as a result of the UN Convention on the Law of the Non-navigational Uses of International Watercourses, the parties have recognised that they have to shift their water disputes from contests of power to considerations of fair rights and mutual obligations. This is because the convention is a comprehensive global code of law that will sooner or later enter into force to govern the management of internationally shared water resources, and inherent in its rules is the responsibility of each state to use water efficiently and to avoid depriving or damaging a coriparian state. These considerations effectively undermine the likelihood of military conflict between Turkey, Iraq, and Syria over water issues and nullify the fanciful scenarios of water war in this basin posed by writers such as Chesnoff (1988). We must now turn our attention to the Arabian Peninsula, in which hydropolitics has a very different setting, not least because the scarcity of water is a major fact of life.

Notes

1 See also: Hardan (1993), Tomanbay (1993), and Wakil (1993). For a detailed description of the physiography and the hydrological features of the Euphrates–Tigris basin see: Naff and Matson (1984: 83–8), Kolars (1991), Hillel (1994: 92–110), Kliot (1994: 100–16), Murakami (1995: 34–42).
2 Literally, Mesopotamia means 'the land between the rivers'. Geographically, it refers to the catchment area of the Euphrates and Tigris rivers which includes southeastern parts of modern Turkey, east and northeast of Syria and the major part of Iraqi territory.
3 Regarding river transportation, it is said that the Greek historian Herodotus was impressed by the amount of commercial traffic in the Tigris River

162 Water Politics in the Middle East

- between Armenia in Anatolia and Babylon (Rzoska 1980). As for urban development, for instance, in 700 BCE. Assyrians constructed several kilometres of canals and aqueducts to carry the Tigris water to their capital city, Ninevah, both for domestic and agricultural purposes (Hardan 1993: 73).

4 However, according to Watson (1983), the Syrian irrigated area increased from 175 000 ha in 1947 to 390 000 ha in 1962, of which 322 000 ha were in the Euphrates Valley, consuming about 4 bcm of water per year. This is another example of data anomaly.

5 Wilcox's report included also the Habbanya project on the Euphrates, the Tharthar project which connects the Euphrates and Tigris, and several project on the Tigris including the Kut Barrage, the Nahravan irrigation project, the Bekhme Dam, and the Mosul Dam (Naff 1991: 4; Hardan 1993: 74; Kliot 1994: 117).

6 For this and other treaties which are mentioned in this section see *United Nations Compilation of Treaties*, NU/S. LEG/SER. B.12.

7 The Koveik (Queik or Qweik) River is a small river which rises in Turkey and crosses the Syrian border. It was previously the source of water for Aleppo in northern Syria. However, because of upstream diversions in both Turkey and Syria, its waters are no longer sufficient for that purpose. The city of Aleppo is currently supplied by Euphrates waters (Kolars and Mitchell 1991: 110, 161).

8 The Hatay territory, according to Kliot (1994: 165), was handed over by France, the mandatory ruler of Syria, to Turkey in 1939 'as a bribe for entering the Second World War on the side of the Allies'. Syria never approved this territorial change and the dispute is yet to be solved. As for the Kurdish insurgency issue, Ankara accuses Syria of supporting the Kurdistan Workers' Party (PKK) in its 11-year separatist war in southeastern Turkey. Damascus scarcely hides its backing for the PKK as a bargaining chip (Barham 1996: 3).

9 This convention was concluded within the United Nations Economic Commission for Europe (ECE), signed at Helsinki on 17 March 1992, and entered into force in late 1996. There is also a valuable work by the International Law Association (1967) which was published as: *Helsinki rules on the uses of the waters of international rivers*.

10 The work of the International Law Commission initially appeared in a provisional draft and was finally adopted by the United Nations General Assembly on 21 May 1997 as the 'Convention on the law of the Non-navigational Uses of International Watercourses', with a vote of 103 in favour, 3 against, and 27 abstentions (see Chapter 2).

11 Because of technical and geological obstacles, construction of the Keban Dam was delayed until 1974 when its reservoir was filled and power production began (Kolars and Mitchell 1991: 26–7; Hillel 1994: 105).

12 This page can be consulted at: http://www.mfa.gov.tr/GRUPC/gap.htm.

13 However, according to Hillel (1994: 105), 'The GAP plan calls for the construction of 80 dams, 66 hydroelectric power stations with a total capacity of 7700 megawatts, and 68 irrigation projects covering up to 2 million hectares.'

14 With no irrigation, the Ataturk hydroelectric power plant would have

a firm discharge of 677 m³/s, firm annual energy production of 8190 GWh, and possible total annual production of 8705 GWh. Under full irrigation, firm discharge would be 375 m³/s, and firm annual production 4550 GWh (*GAP Master Plan*, vol. 2, table 5.4).

15 This point of view was clearly explained by Turkish diplomats to the Sixth Committee of the UN General Assembly in 5 July 1996 (UN General Assembly, 1996, doc. no. GENERAL A/51/2756).

16 It was also in contrast to the definition suggested by the Helsinki Convention (Kliot 1994: 163) and by the Permanent Court of International Justice, according to which an international river is a watercourse that separates or crosses the territories of several states (Chalabi and Majzoub 1995: 219-20).

17 The Orontes river rises in Lebanon and flows through Syria into the Hatay region of Turkey. Its natural discharge at the Lebanese-Syrian border is estimated at 410 mcm, and 1200 mcm at the Syrian-Turkish border. Ninety per cent of the average annual discharge of the Orontes is used by Syria to irrigate its best farm lands, and to provide both drinking water and electric power for western Syria, the country's most populous region. During the summer the river almost dries up before it reaches Turkey. (Beschorner 1992: 29; Bulloch and Darwish 1993: 69; Cooley 1984: 4).

18 As El Morr (1995: 296) points out, the right of each riparian to an equitable share cannot be determined by a simple mathematical formula. The principle necessitates that all interests at stake should be harmonised. Biswas (1993: 21) draws attention to the complexity of the relation between the principle of equitable utilisation and the principle of obligation not to cause harm. However, Bilen (1994: 113) argues that there is always a way to handle this complexity by means of well-considered technical approaches.

19 At the current rate of funding and investment, it could take the Turks more than 50 years to complete the total program (Starr 1991: 29; Mooradian 1992).

20 In 1964 Turkey pledged to release 350 cm/s from the Euphrates but in 1976 agreed to a minimum of 450 cm/s (Tekeli 1990: 210).

21 For detail see *Independent*, 1992 (19 July): 17, (1 August): 13, (31 August): 9, (7 May): 11, (29 December): 8; *TIME*, 29 March 1993: 32; *Geographical Magazine*, 65 (July 93): 10-14; Bulloch and Darwish 1993: 136-140; Hillel 1994: 100-2).

6
Water Politics in the Arabian Peninsula

Introduction

The most paradoxical of the three case-studies is that of the Arabian Peninsula. While this hyper-arid part of the Middle East is the most water-poor and sparsely inhabited region of the world, with the highest population and urbanisation growth rates, there has been less anxiety among neighbours about water conflict and there has been no incident that could be defined as an example of 'water war'. However, Ewan Anderson (1991a: 11) claims that 'the Arabian Peninsula, with a projected annual rainfall of one third the accepted minimum, is an area where conflict over water is likely to arise'. Similarly, George Joffe (1993: 65) predicts that 'during the next decade, water supply may prove to be the region's most politically divisive issue'.

As we shall see in this chapter, the hydropolitical circumstances and water management policies in the Arabian Peninsula throw considerable doubt on the predictions of Anderson and Joffe. Indeed, the Persian Gulf War of 1990–1 and its consequences verify that oil, the dominant strategic resource of the Middle East, will continue to overshadow all other natural resource problems in the region. The truth is that, if Kuwait had been the water tower of the region but devoid of oil resources, it would neither have been subject to invasion nor would its occupation by Iraqi forces have invoked such a considerable reaction from the international community. The Kuwait Crisis showed that Starr and Stoll's (1988: 1) prediction was wide of the mark when they claimed that, 'Despite the expected growth in western dependence upon the [Persian] Gulf oil toward the end of this century, it is safely predicted that water will increasingly shape the future politics of the area.'

We argue, then, that, despite the severe aridity of the region, water is still a resource of much inferior immediate significance to oil in the Arabian Peninsula. This is not to deny that water plays a critical and limiting role in maintaining current agricultural policies aimed at food self-sufficiency or food security. Indeed, as a result of examining the evolution of hydropolitics in the Arabian Peninsula, we suggest that water demand is a fluid concept which can be dramatically affected by small changes in both technology and lifestyle. Likewise, as new developments in Saudi Arabia indicate, attitudes to water and its consumption patterns can change in the face of crisis. We also assert that in order to link the water resources of a region to the wider concerns of security and conflict, we need to consider the underlying cultural, social, economic, and political factors within the region – a point which many analysts who have talked about water war in the Middle East have ignored (see Chapter 7).

By an analysis of the relationship between water availability, as a primordial factor, and the rise of early civilisations in the Arabian Peninsula, this chapter explains that from antiquity, people in this region have seen the scarcity of water as a natural environmental phenomenon rather than a political issue. They have also established proper social and legal institutions to manage their water resources in a sustainable manner. These findings, in turn, reinforce the hypothesis that, if the inhabitants of a region see themselves as confronting a common threat of water scarcity which is resolvable only through their cooperation, then a positive interaction occurs – one which is far from the conventional image of hostility and instability. By examining the latest developments in the region concerning the water management policies of the peninsula's countries, we will see that the hypothesis put forward by Myers (1986: 252), that 'in the Middle East, water is a major factor in political confrontation', is not applicable to the countries of the Arabian Peninsula (Map 6.1).

It is true that some states within the Arabian Peninsula are facing a new challenge to their water resources – affluence. When the meagreness of the peninsula's barren deserts was enriched by the discovery of vast petroleum reserves, these territories entered an age of rapid socio-economic and technological development which drastically

Map 6.1 Countries of the Arabian Peninsula

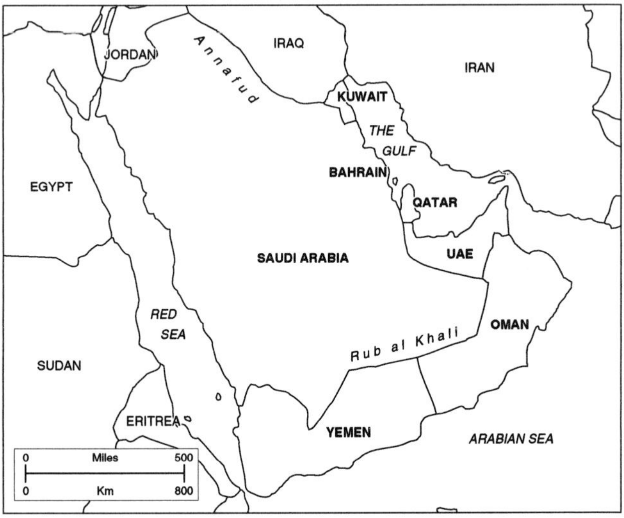

Source: Al-Alawi and Abdulrazzak (1994: 182)

changed the Arabs' traditional lifestyle. The unprecedented wealth provided by oil revenues rapidly increased their standard of living and their population growth rate, which was a function of both natural growth and massive inward migration. The availability of heavily subsidised modern water-pumping and irrigation technology promoted the settlement of the nomadic population and stimulated the migration of the necessary labour force from remote population centres. Another phenomenon that contributed to water resource depletion was the rapid urbanisation in centres located along the coastal zone, where water supplies are limited and quality is continually deteriorating due to salt water intrusion.

All these developments produced a dramatic change in the lifestyle in the region with effects on water consumption patterns that were beyond the capacity of traditional regulations and institutions to sustain. However, the failure of established norms and institutions to manage these ever-increasing water demands led to a water crisis that relates fundamentally to the methods of water allocation and use *within* the countries of the Arabian Peninsula rather than to water allocation *between* them. Moreover, as in ancient times, the very importance of water makes both internal and external cooperation over water more likely than conflict. As we shall show, these recent development in the peninsula indicate that the shared need for optimum management of this vital resource can become a source of regional accord and integration rather than conflict and dissension. But first, we must briefly review the climatic and hydrological configuration of the region.

Climatic and hydropolitical features

Unlike other water basins in the Middle East that are nurtured by bountiful rivers, the great desert peninsula of Arabia is devoid of perennial rivers and streams or lakes. The peninsula is essentially a vast plateau, with an area of about 3 million km², rimmed on the west and south by mountains and sloping gently east to the Gulf. It contains some of the hottest and largest sand dune desert areas of the world, notably the *Rub al Khali* (Empty Quarter) in the south and *an-Nafud* in the north (Encarta 1994). Indeed, more than a quarter of the peninsula's landscape is covered by sand ridges hundreds of miles long and dunes high enough to be called sand mountains. This area is virtually uninhabited because the climate is extremely arid and summer temperatures reach as high as 54° C (130°F).

The countries of the peninsula share a very severe desert environment in which vegetation of every sort, except highly adapted desert micro plants, relies upon groundwater raised from underground aquifers. The climate is generally characterised by high temperatures, high evaporation, low relative humidity, and very low and irregular annual rainfall – between 70 and 130 mm/year. Few areas receive over 100 mm average rainfall, and many receive none at all for months or even years, while potential evaporation rates of more than 3000 mm/year create an impossible condition for a perennial surface water system to exist (Al-Alawi and Abdulrazzak 1994: 174; Al-Zubari 1996: 2).[1] The notable exceptions are the Asir Mountains and the volcanic highlands of Yemen in the southwest of the peninsula and the Oman Mountains to the east. Indeed, Yemen is the most favoured area of the peninsula, in climatic and agricultural terms, and has therefore been known from antiquity as *'Arabia Felix'* ('Happy Arabia') (Grolier 1993). In these areas the physical variations in terms of low and high lands affect the degree of moisture and the amount of precipitation. In both locations rainfall is noticeably higher than elsewhere in the peninsula, thus leading to occasional runoff.

These climatic conditions have a direct impact on the hydrology of the peninsula, which has traditionally been related to water scarcity. This has been the conventional pattern for centuries, during which Arabs have relied on drawing water from the ground. Accordingly, apart from the limited quantities of runoff resulting from irregular flash floods, the most notable *renewable* water resources consist of a system of shallow aquifers some of which originate outside the peninsula as surface flows in the Euphrates–Tigris system infiltrate into the adjacent flood plains, while others accumulate within the interior where there is high precipitation and mountainous relief: Such areas include the Oman ranges, the Asir Mountains, and the highlands of Yemen. These alluvial aquifers have been exploited for centuries throughout the inhabitable districts of the peninsula and other parts of the Middle East (see Burdon 1982). These shallow aquifers are generally thicker and wider in the eastern parts of the peninsula, particularly in the United Arab Emirates (UAE) and Oman, than in the west. Their combined reserves are estimated at 131 bcm; the largest reserves are in Saudi Arabia, with an assessed capacity of 84 bcm (Khordagui 1996: 4). In some areas, the contribution of wadi[2] systems is also of considerable importance.

There is also a massive body of *fossil* water trapped in a series of extensive deep aquifers that hold water hundreds or thousands of years old. These types of aquifers are of regional importance, underlying most of the peninsula and shared by several countries (see Map 6.2). At the beginning of the twentieth century, the search for water above these aquifers led to the exploration of oil in countries such as Kuwait, Qatar, and Bahrain, and, in turn, the widespread search for oil revealed that there are vast reservoirs of groundwater throughout the Arabian Peninsula and Northern Africa (Bulloch and Darwish 1993: 25; Allan 1994a: 82). According to de Jong (1989: 504), these deep sedimentary formations have their outcrop zones in eastern or Central Saudi Arabia and in Southern Oman and, dipping towards the Northeast, they extend partially under the Gulf and the neighbouring states of the UAE, Qatar, Bahrain, Kuwait and Iraq.

The question of whether some of these deep aquifers are being recharged has not yet been resolved. This is because the techniques and equipment necessary to detect, with an acceptable degree of precision, these subsurface conduits are not yet available. However, the popular belief is that fossil aquifers, like oil reserves, are essentially non-renewable. This means that 'pumping water from them

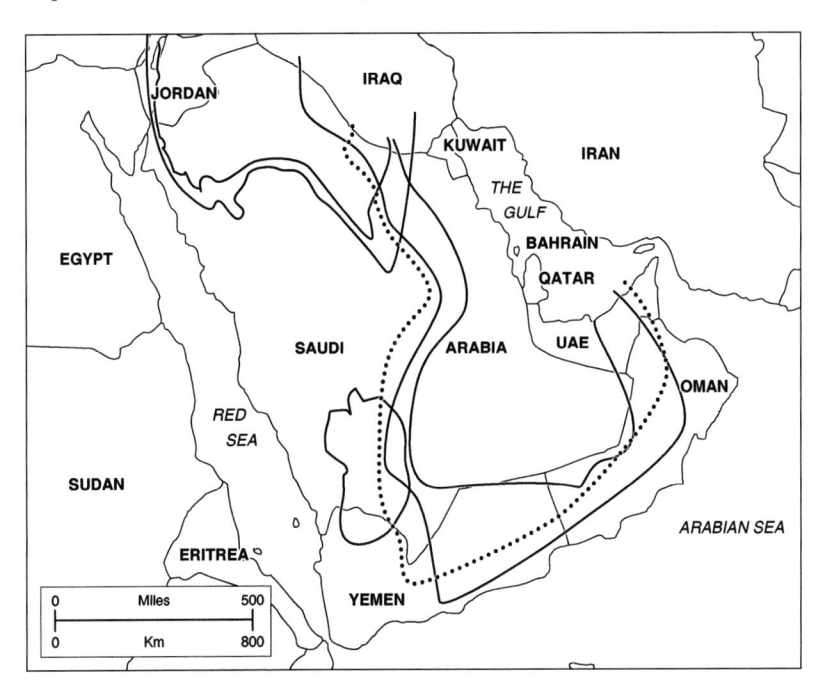

Map 6.2 Shared international aquifers of the Arabian Peninsula

Source: Al-Alawi and Abdulrazzak (1994: 182)

depletes the supply in the same way that extraction from an oil well does. Farms and cities that depend on this water will eventually face the problem of what to do when the wells run dry' (Postel 1992: 31).

Although the predominance of internationally shared groundwater resources is the striking hydrogeological feature of the Arabian Peninsula, the natural water resources are so inadequate in terms of both quality and quality that they cannot possibly close the increasing gap between demand and supply. To escape this predicament, many countries in this region have invested heavily in desalination plants. The countries of the peninsula have uniquely developed on a large scale, non-conventional water supply systems such as sea water and brackish water desalination. Indeed, these countries are pioneers of desalinated water; more than half of the world's desalinated capacity is located in the littoral countries of the Gulf, particularly in Saudi Arabia, Kuwait, and UAE (see below, pp. 189–92).

Table 6.1 Water balance in the Arabian Peninsula

	Annual rainfall (mm)	Annual evaporation[a] (mm)	Annual runoff (mcm)	Renewable groundwater (mcm)	Groundwater use[b] (mcm)	Desalinated water (mcm)	Wastewater re-use (mcm)
Bahrain	70	1 650–2 050	0.20	100	160	56	32
Kuwait	70	1 900–3 500	0.10	–	80	240	83
Oman	71	1 900–3 000	918	550	645	32	10.5
Qatar	67	2 000–2 700	1.35	45	144	83	23
S. Arabia	75	3 500–4 500	2 230	3 850	14 430	795	217
UAE	89	3 900–4 050	150	125	900	342	62
Yemen	122	1 900–3 500	3 500	1 550	1 200	9	6

Source: Al-Alawi and Abdulrazzak (1994: 177)

Notes: a. These figures indicate the potential evaporation rate in the region, which is very much above the real figure
b. The real amount of extraction is subject to wide controversy (see Table 6.3)

Also, in recent years, waste water reuse has gained a high profile in the water policy of the concerned countries (see Table 6.1).

The hydropolitical configuration of the Arabian Peninsula is much more intricate than the other water basins of the Middle East. This is because streams and rivers are easier to locate, regulate, and control than underground waters. As Hillel (1994: 275) points out, the notion of a river basin is easy to grasp, because not only rivers, their tributaries, and their distributaries are visible on the ground and are drawn on maps, but also their flow rates and directions are easily measurable. By contrast, the notion of a subterranean water basin is difficult to conceive, as aquifers are invisible and their water resources are hard to quantify. Moreover, their recharge rates, flow direction, and water quality are always subject to controversy.[3] Like rivers, however, the subsurface water basins are international in character. Indeed, as Map 6.2 illustrates, most major aquifers underlie the territories of two or more countries and there appears to be subsurface transfer of water from the deeper, highly pressurised aquifers to adjacent strata (Gischler 1979: 71). Referring to the facts that groundwater typically ignores political boundaries and that withdrawal by one country can drain the shared aquifer(s) and thereby deprive other neighbouring countries, Ewan Anderson (1991a: 11) and Hillel (1994: 275) argue that there is a risk of escalating conflict between the neighouring states in the Arabian Peninsula. However, as we shall see, the general trend in the region indicates that this is an exaggerated danger.

Evolution of water management in the Arabian Peninsula

Ancient Arabia

Despite severe environmental conditions, Ancient Arabia was a centre of civilisation and a crossroad of trade, and, from time to time, Arab kingdoms arose on the fringes of the desert. However, most of the peninsula became the territory of a highly structured nomadic tribal society (see Hitti 1970). According to Charles Smith (1993: 1), references to Arabs as nomads and camel herders of Northern Arabia appear in Assyrian inscriptions of the ninth century BCE. The name was subsequently applied to all inhabitants of the Arabian Peninsula who were also well known as warriors and controllers of the caravan routes which linked the Indian Ocean with the Mediterranean and served them as an important source of income (Encarta 1994). Indeed, until the mid-nineteenth century, vast semi-desert areas in the Middle East and North Africa were inhabited by Bedouin tribes – Arabic *Badawi*, 'dwellers in the desert' – who lived by herding camels and goats, or by trading. In more favourable grazing areas and oases, Bedouins were associated with sheep and cattle raising activities and limited agriculture. Major peasant populations were settled in the Nile basin and the Mesopotamian Plain both of which were blessed with abundance of water.

The Bedouin tribes, who suffered from years of drought, crossed and re-crossed the parched deserts of the Arabia in search of food and water, sometimes more than 2000 kilometres per year (Spooner 1993: 1). In the cooler, rainy season, they typically migrated in small groups into the desert with their animals. In the hot, dry season they congregated in larger groups around water sources on the desert margins, especially in the vicinity of fertile oases, many of which were the sites of towns and markets. For centuries the history of Arabia centred on the wells and oases as the Bedouin tribes followed the vegetation with their herds, and traders travelled from well to well to obtain water and provisions. In this land, water was the essence of life – indeed life itself – and rainfall was not only the symbolic manifestation of God's power but also the evidence of his grace and mercy (see Chapter 1). Accordingly, water was revered as a gift of God which belongs to the community, and its waste or pollution was considered both a sin and a blatant social offence. Evidently, in such an environment regulating water utilisation was an overriding social organisational task.

The cultural values and social customs shaped by the needs of Bedouins in the desert provided a set of customary rules pertaining to water rights and water consumption patterns. Water resources were regulated by collective interests rather than given over to unbridled private ownership and the patterns were user-focused, intending to influence the users' behaviour, to reallocate existing supplies, to encourage more efficient use, and to promote more equitable access. Historical evidence indicates that traditional rules that grew out of the customs of the desert reflected a sense of communal responsibility regarding water rights and its unpkeep, and that the Bedouin tribes maintained a strict order of priority in the use of a well or a spring – priority was attributed first to human needs, then to animals, and the surplus could be used for irrigation. For instance, a thirsty man could drink from another man's well; he could also lower a container in the well, and the water in the container would become his property; but he should not pollute the well, otherwise he would be punished severely. Since these rules were founded on values that grew out of religious and social customs, they were often more rigid than authoritative state-made laws (Bulloch and Darwish 1993: 160–2).

With the advent of Islam, the greatest event in the history of Arabia, an evolving set of regulations complemented the customs which even sceptical observers such as Bulloch and Darwish (1993: 161) and Hillel (1994: 267) acknowledge that for centuries succeeded, in general, in organising the peaceful sharing of water resources. Born in the desert of the Arabian Peninsula, Islam has been deeply concerned with water both in economic and ritualistic senses. As mentioned in Chapter 1, the connection between the *shari'a* as the 'law of water' and as a generic term for 'Islamic law' was not a matter of coincidence. It was a deliberate choice that illustrates the centrality of water of the Arabs – a centrality which has continued from pre-Islamic times right down to the present day. The *shari'a* developed and codified, in answer to this centrality, an extensive and highly sophisticated system of rules in order to improve water management and minimise conflict (Hillel 1994: 267). The main principles of Islamic water law can be summarised as follows.

1 Water is a gift of God, a *mubah* (free) commodity which should not be owned or controlled to the point of excluding others from using it.
2 The efforts to develop water resources through storage building, distribution facilities, and/or conservation works create a value

added to water which may result in a qualified right to ownership.
3 The right of prior appropriation, combined with the required distribution of surplus, is acknowledged.
4 People are liable for withholding or misusing water, including pollution and degradation of water resources. (See C. Mallat 1995: 129–30)

Regarding the first principle, it should be noted that the shared use of water was traditionally limited to drinking and watering animals, the symbolic purifying of the body for prayer, and to other household requirements. And since there was always enough water to satisfy these domestic needs, this principle eradicated the possibility of conflict over water in this important aspect of life. As for the second principle, the general view was that it is impossible to own water unless it entails a value added by labour. Retaining water in cisterns was a common traditional activity in the desert areas of the Middle East and the trade of water by *Saqqa'in* (street traders of water) is still known in some Arab countries like Saudi Arabia, especially in the pilgrimage season. This principle was particularly encouraging for the development of water resources and land reclamation activities, since the *shari'a* rules on use and ownership of water and land are generally linked together – undeveloped land and water resources are commons that equally belong to the community but could be appropriated through labour. Concerning the prior right of use, water-sharing patterns have been varied according to local uses and prevailing situations. But the general Islamic view acknowledges the right of prior appropriation, while emphasising that the surplus water should be available to those in need with no more extra charge than the costs of its provision and distribution. The literature of Islamic jurisprudence contains frequent cases of *fatwas* (religious edicts) regarding the establishment of prior water rights. These cases are in the form of specific and detailed questions posed to religious sages, followed by their reasoned judgements (Caponera 1973; see Hillel 1994: 267–9). The interesting point is that, according to Hillel (1994: 268), the edicts issued by medieval Jewish rabbis on water right settlements were generally similar to those of their Islamic contemporaries. This historical fact not only highlights the role of the natural environment in the constitution of social norms in the region, but also its effect on the formation of cultural and religious rules and values. This important point will be discussed in the concluding chapter.

To prevent the monopoly or misuse of water resources, a detailed classification of wells – the major water source in the region – was presented under the *shari'a* rules. Wells were divided into three categories. The first category consists of the wells which are dug on public highways to serve travellers and their animals. Everyone has free access to these wells and no one, including those who dug it, can claim ownership or to determine the way people utilise it. The second group includes the wells which are dug to be used for a particular purpose in a specific period. When the specified purpose is fulfilled or the announced period is over, the well becomes a public asset and those who dug it can benefit from its water like everyone else. The third category belongs to private wells which are on private land (Bulloch and Darwish 1993: 175). Though the owner of a private well has his own priorities, he must allow his neighbours a share of the well's water for their domestic needs (Hillel 1994: 267). To prevent the degradation of underground water resources, wells are protected by regulations concerning the mandatory distances that should be observed for digging additional wells in the vicinity of an existing well or siting sewage wells in its nearby areas. The regulations vary according to both the hydrological and geological circumstances of the concerned area.

Applying the same criteria to springs, the *shari'a* divided them into three categories: the first was the natural fountains whose running waters gush through natural forces. The access to these waters is *mubah* or 'free for all people' as long as there is enough water to meet everybody's requirements. Otherwise, those who live in its immediate vicinity – a village or a tribe – have a communal protected right of prior use. The second category includes the springs which are developed and caused to flow by human deed. The right of joint ownership and common access to these waters belongs to those who participated in its development. If there is surplus water, those who live in the vicinity have the right to benefit from the spring. Otherwise, priority is given to the owners. If water is not plentiful enough to meet all their needs, then the owners will apportion water according to their share of contribution to the project and to its maintenance. The third category belongs to springs uncovered and developed on private lands. Here the right of private ownership is recognized. Nevertheless, the owner should let others take it who need his surplus water (Bulloch and Darwish 1993: 174–5; Chibli Mallat 1995: 131).

As far as Bedouins were concerned, the common uses of available water resources were adequate to satisfy their modest needs. However, the historical accounts indicate that whenever there has been a war in the desert area, water has become a strategic object to be denied to an enemy. For instance, Drower is quoted as saying that, in the sixth century BCE, Assurbanipal, King of Assyria, seized water wells as part of his strategy of desert warfare against Arabia (Gleick 1993c: 109). Such incidents, however, should not be regarded as examples of water-imperative war, since water was not the cause of war in the first place. In contrast to the pastoral Bedouins, in areas where sedentary populations were engaged in agricultural activities, the principle of common use was not adequate to meet all water requirements of the community. By introducing the second principle – the notion of ownership of land and water through labour – Islam encouraged both the development of water resources and the revival of dead land. Indeed, the reclamation of barren land and the ownership of its underground water were two sides of one coin.

Along with the institutionalisation of water rights and the establishment of water sharing procedures, which effectively minimised the possibility of dispute over water, the inhabitants of the Arabian Peninsula practised innovative water management methods to overcome the aridity of their environment and to sustain a subsistent economy. Among the techniques which were developed, three have survived: water harvesting, spate irrigation, and *qanat* systems (Agnew and Anderson 1992).

Besides some mechanical lifting devices which raise water from wells, since antiquity, water harvesting has been a simple method of water supply management which has been prevalent at local level throughout the arid parts of the Middle East. Although it is still a common practice in countries of the region, it is more suitable for satisfying very modest local needs and directly depends on rainfall. Today water harvesting is mainly used for groundwater recharge (see next section). Spate irrigation technique has been successfully applied in southern parts of the Arabian Peninsula for centuries and still is regarded as an important method of water management in Yemen, where flash floods normally occur (Saif 1987). This method is not only used for irrigation but also for recharging underground aquifers which satisfy the domestic water requirements of the country. The *qanat* system is another effective ancient water supply system in the Middle East. In eastern parts of the peninsula, notably in Oman and to some extend in Yemen, *qanats* have functioned successfully for more than 2000 years (Agnew 1995: 26). These remarkably ingenious and sustainable systems served, for millennia, as the principal means for supplying water to towns, villages, and orchards (Hillel 1994: 16).

One striking feature of these schemes is that they are very labour-intensive; both their construction and maintenance depend heavily on the availability of manpower. For instance, maintaining spate irrigation schemes requires a highly organised, coordinated, and skilled labour force to build, shift, knock down and alter small earth dams to direct floods to the cultivated farmlands (Bulloch and Darwish 1993: 195). The *qanat* systems need even more organised efforts to be built and to be kept operational. As Figure 6.1 illustrates, a *qanat* is a human-made spring consisting of a 'chain-well' system with one or more mother wells which drain laterally through an underground conduit that carries water some distance to emerge at ground surface – often many kilometres away from the mother well. The vertical shafts are used to dig the system and to keep water running in the horizontal tunnel, which starts in a shallow aquifer and slopes to intersect with the underground water table and ends at ground level where water is used for irrigation or domestic purposes. Since much of the network runs underground, the *qanat* systems proved highly effective in keeping the water clean and preventing evaporation – a striking feature of the arid climate. Another major advantage of the *qanat* system is that the ground water cannot be over-abstracted, so a moderate sustainable supply of water is available all year round. It is economical too, because there is no need for energy to pump water or to force it along. However, there is a major disadvantage in that the amount of water running through the system is totally dependent on the level of the water table, which in turn depends on the natural recharge of the aquifer.

Moreover, these long systems, though constructed with simple technology and with materials locally available, need constant attention and require a well-established cooperative mechanism among the community who live in its vicinity. For instance, additional wells must be dug sufficiently apart to avoid mutual interference. Furthermore, maintenance of the *qanat* system necessitates an organized division of labour and adequate skilled labour force, while its water management needs an established right of ownership and a headman or manager whose principal task is to allocate a schedule of times when each villager can open the sluice of his field or garden to the *qanat*. Such rights become part of local inheritance

Figure 6.1 The structure of a *qanat* system

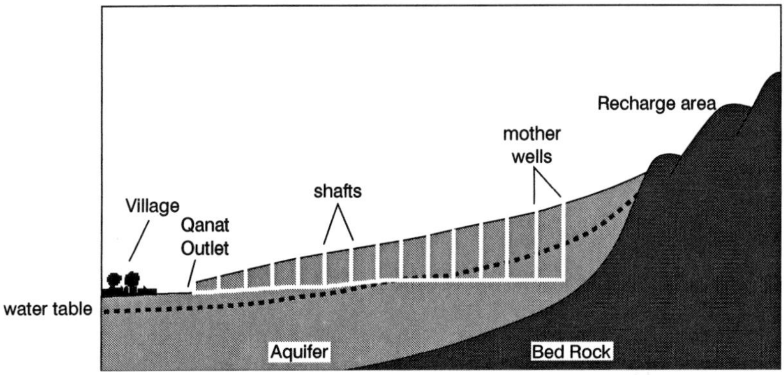

and the headman needs enormous patience to sort out the occasional disputes.

Until modern times, several cities and villages in the Middle East and North Africa, from Morocco and Algeria to Iran and down the Arabian Peninsula to Oman and Yemen depended on *qanats* for their water supply. According to Bulloch and Darwish (1993: 194), in Oman and Yemen, the *qanat* system is still the main network upon which local farmers rely for irrigation – the existing system supplies and irrigates 71 per cent of the cropped area in Oman. According to Bodgener (1984: 38), archaeological studies indicate that these ingeniously constructed systems served cities, villages, and irrigated fields in the southern and eastern parts of the Arabian Peninsula as long ago as the third century BCE. These systems also nurtured a culture of cooperation and communal identity in the region. Thus it is not just for their efficiency that there are now pressing calls for these aqueducts to be preserved in the face of modernity, but also because they are reminders of a cultural heritage.

However, despite their efficiency and sustainability, these traditional methods of water management could no longer satisfy the ever increasing water requirements in the Arabian Peninsula. Since the 1950s, and particularly since the 1970s, a great strain has been placed upon these sources as the pace of oil-driven modernisation has steadily increased. With the arrival of motor pumps and modern farming technology, over-extraction of shallow aquifers has led to the constant lowering of the water table and, inevitably, to the drying up of wells, springs, and *qanats*.

The post-oil era not only disrupted the traditional water management systems in the peninsula, but also brought about numerous changes in the political, economic, and social aspects of life in the region: the nomadic tribes acquired flags and national boundaries; the wealth of oil changed the tribal subsistence economy to a lucrative consumer-oriented market economy; the values shaped by the ideas of modern farmers and rich urban dwellers contradicted those formed by the needs of nomads in the desert; and the water supply systems that effectively provided water for villages and tribes for centuries failed to do so when the tribes became sovereign nations with a burgeoning urban population whose demand for food and water was steadily increasing.

However, the traditional civil institutions that governed water sharing between tribes and villages remained intact. Since water was still regarded as a gift from God and, hence, a *mubah* or 'free' property that belongs to the community and to which a value could only be added by labour, the possession and administration of water resources was considered as an original right and a duty of states. Such a perception of water imposed the responsibility of provision, management, and distribution of water on governments. Before analysing the response of the governments to this delicate responsibility, in the next section, we will examine the main factors which disrupted the traditional water management systems in the peninsula and the increasing magnitude of water scarcity in modern Arabia. An account of recent developments in this extremely parched corner of the world – mainly rapid population growth, unprecedented urbanisation, progressive rise of living standards and social expectations, and the sudden change of a barely subsistence economy into a consumptive economy – is a prerequisite for an understanding of the theory of future water conflict in the Arabian Peninsula.

Modern Arabia

From what has been discussed so far, it should be clear that until the twentieth century, available supplies of water were commonly used to meet the domestic needs of the inhabitants of the Arabian Peninsula and to sustain the very primitive lifestyle of the Arab tribes. This was not just because the peninsula was sparsely populated, but also because both its nomadic and sedentary inhabitants adjusted their water consumption patterns to conform to the particular specifications and requirements of their natural environment. Living in a desert environment, however, they were always at the mercy of nature: due to recurrent droughts, their tight water supplies were always subject to fluctuation; their grazing herds were prey to starvation; and the people were prone to privation, famine, or outward migration. But from the 1930s when the intact pools of oil were uncovered and exploited in the wasteland of Arabia, this regional stereotype was reversed.[4] The discovery of oil profoundly transformed the peninsula from an impoverished region of small desert principalities to an area of affluent emirates and kingdoms with the highest per capita incomes in the world outside the OECD. With such a tremendous windfall, the peninsula's subsistence economy changed into a lucrative oil-based economy and the oil revenues financed modernisation projects, particularly in health, education and other social improvements. The oil boom of the 1950s and the 1970s accelerated the economic development in the region and established a modern and high standard of living. Such a rapid transition has had several consequences for the countries of the peninsula including change of demography and rapid population growth, fast-growing urbanisation, pollution, food shortage and by extension, water scarcity, and a breach with traditional values that has led to extravagant and unsustainable use of natural resources. As we shall see later, the most remarkable aspect of this rapid transition was not so much the passage from want to affluence as the passage from labour to leisure. But first let us sketch a general outline of these changes and their repercussions on water resources management in the peninsula.

The first important sociological issue that ties closely with economic development in the Arabian Peninsula, one which sharply strikes at the water balance of the region, is demographic change including population growth and urbanisation (Fahim 1996: 5). Regarding demographic change, Allan and Warren note that the proportion of dryland inhabitants following a nomadic lifestyle in the Middle East and North Africa had fallen from more than 20 per cent in the 1950s to less than 3 per cent by 1990 (cited in Agnew 1955: 22). Moreover, as a result of major political, economic, and social changes that have occurred in the Arab world during this century, the whole Arab society today is more heavily urban than rural (Charles Smith 1993: 3). This is particularly outstanding in the countries of the Arabian Peninsula, where the pace of economic development has not only made the Bedouin population more

Table 6.2 Population growth in the Arabian Peninsula (millions)

	1960	1990	2010	2025
Bahrain	0.2	0.5	0.85	1.0
Kuwait	0.3	2.1	3.22	4.4
Oman	0.5	1.5	2.99	4.3
Qatar	< 0.1	0.4	0.64	0.9
Saudi Arabia	4.1	14.1	30.5	44.8
UAE	0.1	1.6	4.87	2.7
Yemen	5.2	10.5	18.98	29.7
Total	~ 10.5	30.7	62.05	87.8

Sources: World Resources Institute (1990: 255); Bureau of the Census, US Dept of Commerce (1994)

sedentary but has also absorbed large numbers of exogenous people into the socio-economic mainstream of the region.

As Table 6.2 indicates, the population in the peninsula's countries has increased almost threefold during the period 1960–90, from 10.5 to 30.7 million. If we disregard Yemen – the *'Arabia Felix'* which since ancient times has been the most fertile and thus most densely populated section of the peninsula – the population growth has been fourfold, with Saudi Arabia having the highest concentrations of population as well as the largest territorial area. According to Mahdi (1996: 7), the projected population level for the year 2000 has already been surpassed in all the peninsula's countries except Bahrain, and the level for 2010 has been exceeded in most of them, since the annual population growth rate in these countries has been far above 3 per cent.

These growth rates partially reflect the impact of modern medicine and social services that have lessened infant mortality and increased life expectancy. They also show that, although traditional tribal lifestyle has nearly disappeared, tribal values have retained their importance (C. Smith 1993: 4). In the case of population, the prevalence of traditional attitudes favouring large families has withstood the tendency towards smaller families found in developed countries. This fact is clearly visible in Saudi Arabia, whose population is now estimated to be over 20 million, up from 3.4 million in 1950. The immigration of an exogenous labour force attracted by the petroleum boom has also been a major contributor to the population growth in the oil rich countries of the Arabian Peninsula. For instance, about 70 per cent of the UAE's population growth in 1992 was accounted for by net inward migration and the remaining 30 per cent was the result of the natural growth of both the native and the migrant populations. As for Qatar, which was a tiny village before oil production began in 1949, only 25 per cent of the population are Qataris; the rest are immigrants from other countries. Some demographers, however, argue that falling birth rates will at some date in the twenty-first century bring the region's population to a low, or even a zero rate of growth (Allan 1994a: 92).

The second most important problem that has contributed to water resource depletion in Modern Arabia is increasing urbanisation, mainly in the form of large urban centres located along the coastal zones in which the beneficiaries of oil revenues live in the lap of luxury. Ironically, in these areas natural fresh water supplies are limited and their quality is continually deteriorating because of salt water intrusion and excessive urban and industrial effluence which contribute to environmental deterioration. The urbanisation rate in the peninsula in 1990 was claimed to be 96 per cent in Kuwait, 90 per cent in Qatar, 85 per cent in Bahrain, 78 per cent in UAE, 77 per cent in Saudi Arabia, 21 per cent in Yemen, and 12 per cent in Oman (Abdulrazzak 1996: 2). This pace of urbanisation and social change is expected to continue well into the next century, because of lack of water, fewer opportunities in agriculture, social modernisation, and young people flocking to major cities seeking education and employment. Increasing urbanisation results in a considerable increase in domestic water requirements and other basic social needs. This, in turn, place serious strains on governmental abilities to respond properly to these needs. Accordingly, some observers, such as Mahdi (1996: 7), have recently questioned the rationale of encouraging the settlement of nomads in these highly urbanised societies.

The drastic growth in population and urbanisation, along with the rise in the standard of living and social expectations, exacerbated the problem of food scarcity in the region. The lack of agricultural productivity forced the Arabs to increase their annual food imports from $1.7 billion in 1972 to $12.7 billion in 1982. In the peninsula, by the early 1990s more than 90 per cent of the Gulf states' food requirements were imported (Starr 1995: 59). However, this heavy reliance upon food imports was viewed not only as a drain of national financial reserves but also as a source of instability and insecurity (Mahdi 1996: 6), and despite severely limited possibilities for further agricultural expansion,[5] for political reasons all the peninsula's countries adopted a policy of food self-sufficiency. Disregarding the fact that a goal such as food security, when defined as self-sufficiency of individual countries or of the Arabian Peninsula as a whole, needs to be evaluated in terms of the resource endowments of the region, the governments provided extensive support to the modern farming sector. Misinterpreting the objective of *food security* as *food self-sufficiency*, modern farming in the post-oil era received the highest allocation of water supplies and thereby drained off the finite water resources of the peninsula in search of an illusory objective.

In order to satisfy their domestic and industrial water requirements and to provide food in the desert, therefore, the countries of the Arabian Peninsula have, during the last twenty years, faced increasing water deficits (see Table 6.3). According to Al-Alawi and Abdulrazzak (1994: 186), water demand in the peninsula rose from 6 bcm in 1980 to 22.5 bcm in 1990, and it is expected to exceed 31 bcm by the end of this century and to more than 35 bcm by the year 2010. Among the region's water users, the agricultural sector is the largest consumer. In Saudi Arabia, for instance, agricultural water use increased eightfold between 1980 and 1990; and across the peninsula all accessible water resources were tapped and exploited far faster than they could be replenished. Regarding the renewable water supplies of the peninsula's countries, their per capita water deficit in irrigation for the year 1991 has been reported as follows: Kuwait 1020, Qatar 1015, Bahrain 1008, UAE 998, Oman 840, Yemen 625, and Saudi Arabia 456 m^3 per capita (Mahdi 1996: 2.3).[6]

These figures indicate that in the Arabian Peninsula, achieving a high degree of food self-sufficiency leads inevitably to a rapid depletion of water resources. Indeed, it has already led to the accumulation of an outstanding annual water deficit in this region (see Table 6.3). However, it is not just the agricultural sector that should take the blame. Extensive use of water to provide green amenity in urban areas, public greenery in parks, landscaping, green belts, and road-side ornamental trees in a desert environment is also a water-intensive activity. The UAE is one of the extreme examples of the extravagant use of expensive water for such objectives (Allan 1994a: 78). Moreover, most of the countries of the region have witnessed a 20 to 30 per cent *annual* increase in their domestic and industrial water demands over the last 10 years (Khordagui 1996: 6). Last but not least, growing pollution from agricultural,

Table 6.3 Water deficit in the Arabian Peninsula (million m³/yr)

	Surface water	Groundwater recharge	Present exploitation	Present deficit
Bahrain	0.2	100–112	190	78–90
Kuwait	0.1	160	114[a]	n.a.
Oman	918	475–550	728	–
Qatar	1.35	45–50	185	133–9
Saudi Arabia	2230–3210	2340–3850	14 430	7370–9860
UAE	150	120	1000	730
Yemen	3500	1550	3400	–
Total	6800–7780	4790–6392	20047	5875–8457[b]

Source: Al-Zubari (1996: 11); Al-Alawi and Abdulrazzak (1994: 177); World Resources Institute (1990: 331)

Notes: a. This does not include water abstracted from private wells in agricultural areas.
b. This excludes Kuwait

municipal, and industrial sources is a considerable factor which takes a heavy toll on already limited water supplies of the region (G. Joffe 1993: 67). This is partly because, despite rapid economic development in the region, most of the countries still lack a clearly defined water pollution control policy, especially in coastal and marine areas (ECWA 1978; Al-Kuwari 1996).

Before we discuss the response of the concerned countries to this unsustainable hydrological situation, it should be noted that the aforementioned socio-economic developments in the Arabian Peninsula exerted most pressure on the aquifers (de Jong 1989: 503) and if this trend continues, they will soon be exhausted. Moreover, these aquifers, which constitute the major water resources of the region, are to a significant degree shared across international boundaries (see Map 6.2) and their sustainability is a matter of concern for all collateral countries. However, a search of international water resources treaties relevant to the Arabian Peninsula reveals that there are no binding legal instruments relating to the utilisation and the joint development of the shared international water resources in this region (FAO 1978). Even after the establishment of the [Persian] Gulf Co-operation Council in May 1981, there has been little progress in this direction and the region still lacks an explicitly binding agreement. This is partly because international groundwater law and principles are poorly developed[7]; and partly because decision-makers in the oil rich countries have presumed that their wealth of energy could be the final and magical negation

Table 6.4 Per capita water availability from natural sources (1990)

	m^3
Bahrain	<10
Kuwait	<10
Oman	1360
Qatar	60
Saudi Arabia	160
United Arab Emirates	190
Yemen	180
Middle East	1250
World	7690

Source: World Resources Institute (1990: 330–1)

of water scarcity (we shall explain this point in the next section). However, despite being the region with the poorest per capita water availability (see Table 6.4) and the highest per capita median cost of water supply and sanitation in the world,[8] and in spite of the absence of any obligatory international water agreement among the peninsula countries, there has been no documented case of violent dispute over water among parties in this region. Indeed, the current hydropolitical situation in the Arabian Peninsula undermines the credibility of the theory of 'water war' and the predictions made by some writers who have suggested that the Arabian Peninsula is an area where conflict over water is likely to rise (see for instance, Starr and Stoll 1988: 6–7; E. Anderson 1991a: 11; Bulloch and Darwish 1993: 24). Having observed that this prediction has not come true, Allan and Mallat (1995: 15) claim that water would have been a source of conflict in the region if the concerned governments had insisted on observing their food self-sufficiency policies. They argue that:

> Those managing Middle Eastern economies have in practice traded rationally and have gained access to very cheap water via international trade in food. With the massive volumes of water being so readily available to meet the needs of the rapidly rising populations it has been possible to provide water for essential domestic, municipal and industrial needs and such provision will also be possible for the foreseeable future.... In these unpressured circumstances it will become increasingly possible to co-operate over access to water and over the allocation and the re-allocation of water. (Allan and Mallat 1995: 15)

However, although this argument is relevant to some Middle Eastern countries, such as Egypt, Israel, and Jordan, it is not wholly applicable to the countries of the Arabian Peninsula. As we shall see in the following section, agriculture is still the main consumer of the peninsula's natural water resources, and domestic, municipal and industrial water requirements are primarily provided by desalination of sea water and brackish water. For these countries, so far, the challenge has been one of meeting the ever-increasing water demands by *engineering* new sources of water. Indeed, they have considered their *energy*, rather than the *imported food*, as a substitute for water. While some countries plan to rely on partial food imports, others aim at achieving self-sufficiency in food production and possibly even to export it. Accordingly, the estimated level of water resources required to satisfy future needs depends entirely on the national strategy of economic development and the standard of living generated by the natural and financial resources of each particular country in the region. Let us see how these countries have responded to their common hydrological predicament – has it been a conflictual, a cooperative, or a unilateral response?

Water management policies of the regional states

What follows from the preceding discussion and the earlier parts of this chapter is a recognition that in the Arabian Peninsula water scarcity is the most constricting factor which adversely affects the process of socio-economic development in the region. However, the water management policies adopted by the peninsula's countries indicate that, despite the severity of the issue, for these societies water is a resource of less immediate significance than oil. Contrary to what western strategists suggest, the dominant perception in the region is that water scarcity is a problem which can be resolved by economic and technological measures. For example, as a result of the Iran–Iraq War and the Persian Gulf War, when the vulnerability of desalination plants to oil pollution induced a sense of water insecurity among those countries which are heavily dependent on these installations, the concerned countries have resorted to long-term technological solutions. Saudi Arabia and Kuwait are both working on plans for storing up to one year's supply of water in the strategic storage facilities in the porous rock aquifers that have been exhausted (Bulloch and Darwish 1993: 195).

The perception of water as an economic and technological issue that was established in the oil era as the decision-makers in the oil rich countries presumed their wealth of energy resources could be the virtual negation of water scarcity, indicated that for them oil was not just a symbol of affluence but of order, of control over the uncontrollable – water. Peninsular governments found that they were no longer at the mercy of nature, because, with money at hand, if nature failed, modern technology could intervene to make up the shortfall. This point may be best illustrated by conditions in Saudi Arabia, a country that traditionally managed to survive on its modest renewable supply of water but, according to (Butler 1995: 38), in recent years has pumped out some of its aquifers at 100 times the sustainable level. Other countries of the peninsula have followed the same line in identifying and exploiting more water resources, depending on their respective capacities in both financial and human resources. Unlike other water basins in the Middle East, there is no point of surplus of the peninsula from which water could be diverted or transferred to the area of distress. All collateral countries, to one degree or another, suffer from lack of regular rain-fall and the absence of permanent surface waters (see pp. 167–70). Having faced a common threat of water scarcity, their response to the threat has been always similar – adopting new means of enhancing the limited water supplies. In the next section, we shall discuss the water management policies adopted by the peninsula countries, and their merits and drawbacks. Their effects on enhancing cooperation or conflict in the region will be examined in the concluding section.

Groundwater management

When the widespread search for oil uncovered the vast reservoirs of groundwater in the Arabian Peninsula, it came as a great relief for the governments, who were already finding difficulty in providing a steady and adequate supply of food and water for their growing populations. Ever since, and especially recently, in their drive for food self-sufficiency, the concerned governments have used modern water-pumping and irrigation technology to transform desert dunes into fertile fields (E. G. H. Joffe 1995; Keen 1988; Duwais 1990). They have also allocated considerable funds and efforts to enhance these important sources of fresh water; for instance, earth dams have been constructed in *wadis* to boost both conservation and recharge of shallow aquifers. About 195 dams of various sizes have been constructed in Saudi Arabia for this purpose with a combined storage capacity of 475 mcm; and 52 dams have been and are being constructed in Yemen, UAE and Oman (Musallam 1990: 16; Khordagui 1996: 4). Moreover, in better off areas of Yemen and Oman enormous amounts of money have been spent to retain the traditional waterworks. For example in its second development plan (1981–5), Oman devoted more than $64 million to the maintenance and repair of its *qanat* systems which depend on shallow aquifers (Bodgener 1984: 38).

However, in the absence of collaboration among the countries sharing common water resources, over-pumping of shallow aquifers has led to a steady decline in the water table and, consequently, to the incursion of saline water from the coast as landward water pressure decreases. According to Starr and Stoll (1988: 4), because of dramatic increases in water withdrawal from these aquifers, the freshwater supplies of Bahrain, which are derived from the massive aquifers under Saudi Arabia, have been entirely destroyed. Similar developments have been taking place in other countries of the region, including the eastern parts of the peninsula, particularly in UAE and Oman, where the shallow aquifers are generally thicker and wider than in the west (Khordagui 1996: 4). For instance, in the Batinah, the most important agricultural area of Oman, there has been a marked increase in salinity along the coastal zones, and, in the interior, over-pumping has resulted in a dramatic decline in the water table beneath Buraimi-Al ayn area, a border zone between Oman and the UAE in which ideal natural recharge conditions have traditionally sustained high-yielding aquifers. During the 1980s, a fall of at least 50 metres was recorded in this area mainly as a result of over-pumping in the UAE (Starr and Stoll 1988: 4, 18–19). As mentioned earlier, many observers have claimed that since major water resources in the Arabian Peninsula are shared by two or more countries, without international agreements between them, the current water policy of continuously increasing pressure on the regional groundwater resources provides an obvious potential for conflict. The same argument is applicable to the deep aquifers of the peninsula (see Map 6.2).

Now, it is technically possible to tap these aquifers at depths as great as 1700 metres (Musallam 1990: 16). Indeed, one of these aquifers, called *Minjour*, has already been drilled to depths of 1200 to 1400 metres to supply a part of the water demand of the Saudi capital, Riyadh (AOAD 1996: 4). As Map 6.2 shows, these aquifers lie mainly under the central and northern parts of Saudi Arabia and to a lesser extent beneath other countries of the region. It is also in Saudi Arabia that the intensive development of agriculture is taking its major toll on these aquifers.[9] According to Ahmad (1996: 4) and Temperley (1992: 19), about 90 per cent of the irrigation water used in Saudi Arabia and throughout the peninsula is pumped from the underground reservoirs, which almost without exception have been considerably reduced both in quality and in static water levels. Most recent studies and reports indicate that if the present rate of withdrawal is maintained, a shortfall in groundwater production will occur within the immediate future (Gischler 1979; Al-Alawi and Abdulrazzak 1994; Raphael and Shaibi 1984; Al-Sheikh 1996; Khatib 1996).[10] Bearing in mind their negligible rate of recharge, Ahmad (1996: 4) argues that these fossil water resources will not last long beyond the beginning of the next century. Assuming that only 80 per cent of these groundwater reserves can be exploited, Postel (1992: 32) believes that at the current rate of extraction the supply would be exhausted in 52 years. Temperley (1992: 19), however, believes that these sources could endure for nearly 100 years, though economically it would be infeasible to use them.

Regardless of how long these aquifers will maintain their viability, it is evident that in the Arabian Peninsula a real water crisis is looming on the horizon. By using their finite oil and gas resources to subsidise ever increasing consumption of their exhaustible water resources (all countries of the peninsula are turning oil into water), they may mask the water scarcity for a while, but there is no doubt that sooner or later both oil and water wells will run dry. Despite the severity of the issue, however, the concerned countries have been reluctant to engage in a serious and explicit debate over their badly needed and rapidly depleting shared water resources. Although the concept of internationally shared water resources was recognized by the Congress of Ministers of Agriculture from the peninsula's countries and the available data on those resources were subsequently surveyed and evaluated (FAO 1979), no clear plan for their coordinated development was agreed (de Jong 1989: 508–9). Instead, every country in the peninsula has unilaterally attempted to identify and develop its own accessible water resources.

Non-conventional water management

In the Arabian Peninsula the inadequacy of natural water resources in terms of both quantity and quality did not, however, provoke hostility and conflict among the neighbouring states. Instead, it led to a concentrated development of non-conventional alternatives to bridge the gap between supply and demand. Large-scale desalination of sea water and brackish water has been the most common alternative adopted in the region since the 1950s. Wastewater reclamation has also been practised, albeit to a limited extent. Moreover, the economic and technological feasibility of other alternatives such as importing water using supertankers to deliver water like oil, towing of large capacity rubber bags (Medusa bags), and interbasin transfer of water from adjacent regions through pipelines has been studied. Even the prospect of less-realistic alternatives such as towing icebergs from the Antarctic to melt down for tap water, weather modification, and the newly developed technique of growing crops on sand using sea water has been contemplated.[11]

Desalination

With ever-increasing depletion of groundwater resources in the Arabian Peninsula, large-scale desalination has been adopted as the foremost practical solution to the water crisis. Since 1958, when Kuwait pioneered massive desalination technology as a major source of water supply, most of the urban centres in the region have switched to desalination of sea water and brackish water to provide for their municipal and industrial water demands.[12] Consequently, the region has become the largest market for manufacturers of desalination equipment. Indeed, during the last three decades more money has been spent on desalination plants in the Arabian Peninsula than in any other part of the world, a reflection of the region's dire shortage of water.

According to data published by the International Desalination Association (Wagnick 1990, 1992), countries of the Arabian Peninsula have become the world's top consumers of desalinated water. And despite their relatively enormous costs, they are going ahead with the construction of more desalination plants. According to Al-Alawi and Abdulrazzak (1994: 183), the present installed capacity of desalination plants in the Arabian Peninsula has reached 2.02 bcm/yr, compared with the worldwide capacity of 5.68 bcm/yr. However, including the privately-owned plants, other accounts suggest that the desalination capacity in the peninsula exceeds 2.6 bcm/yr (see Kolars 1993: 46, Gleick 1994a: 39).[13] As Table 6.1 indicates, Saudi Arabia produces more than 51 per cent of the desalinated water in the peninsula followed by the UAE (22 per cent), Kuwait

(15 per cent), Qatar (5 Per cent), Bahrain (4 per cent), Oman (2 per cent), and Yemen (1 per cent). In 1992, the operational desalination plants were distributed in the region as follow: twenty-three major plants in Saudi Arabia, eight in the UAE, six in Kuwait, three in Bahrain, two each in Qatar and Oman, and one in Yemen (Al-Alawi and Abdulrazzak 1994: 183, Khordagui 1996: 4). The water produced by these plants constituted 51 per cent of the region's domestic and industrial water supply and is expected to increase to 58 per cent by the year 2000 (Khordagui 1996: 14).

Optimists believe that desalination of sea water is the only reliable alternative to depleting ground water resources, arguing that this unlimited resource could meet all the foreseeable domestic and industrial water requirements of the Peninsula and the world in general (Temperley 1992: 27). According to this argument, the significant advantage of resorting to large-scale desalination as a method of water management policy is that it provides a measure of water security. If a sense of water security is achieved, the possibility of conflict would fade.

However, the resort to desalination as a means of water supply management has been criticised on economic, environmental, and security grounds. From an economic point of view, the major disadvantage of this non-conventional method of water management lies in its high cost. Accounting for just two-tenths of one per cent of world water use (Omicinski 1996), desalination is still too expensive to use except in countries like those in the Arabian Peninsula, where oil revenues provide the necessary capital for initial investment, and readily available sources of cheap energy facilitate the maintenance of the energy-intensive desalination plants. Postel (1992: 47) rejects the option of desalination as a sound method of water management, arguing that 'desalination holds out the unrealistic hope of a supply-side solution, which delays the onset of the water efficiency revolution so urgently needed'. However, Khordagui (1996: 14–15) argues that under present circumstances desalination is one of the most cost-effective solutions to water scarcity in the Arabian Peninsula. This argument is more applicable to desalination of brackish water. Being too salty to drink but much less salty than sea water, brackish water costs less than half as much as sea water to be desalted – currently at 40–70¢ per m^3 compared to sea water at \$1–2 per m^3 (Postel 1992: 45–6).[14]

From an environmental perspective, however, desalination cannot provide the answer because its high energy use runs counter to the goals of reducing air pollution, acid rain, greenhouse gas emissions, and fossil fuel conservation (Starr 1995: 191). In response to this drawback, those countries which are heavily dependent on desalination technology as a means of water supply management have seriously considered the development of solar energy as a substitute for fossil fuel. For instance, from the early 1980s in Saudi Arabia and the UAE experimental work has been carried out on solar-powered desalination projects (see Charnock 1984: 28, El-Nashar and Qamhiyeh 1991). Extensive research has also been conducted at the Kuwait Institute of Scientific Research, and the Qatar Solar Energy Research Station (Starr and Stoll 1988: 4). Considering the climatic conditions, the use of solar energy for desalination in the Arabian Peninsula could be the most economically feasible and environmentally sensible solution provided that solar technology proved to be reliable. According to El-Nashar and Qamhiyeh (1991), the five-year operating results of the Abu Dhabi Solar Desalination Plant, 1985-9, indicate that the plant had a reasonably good record of reliability and ease of operation and the solar collectors were trouble-free and exhibited only a slight drop in performance. For these countries the development of solar energy is a promising alternative which could avert disaster when the oil wells run dry.

From a security point of view, the main disadvantage is that desalination plants are extremely vulnerable to both accidental and intentional damage. The Iran-Iraq War (1980-8) and the Gulf War (1990-1) demonstrated that desalination plants can easily be attacked and are seriously threatened by off-shore oil pollution; in fact, if only a trace of oil enters the inlets of a desalination plant, it can cause considerable damage that would take significant time and effort to repair (Musallam 1990: 15). During the Iran-Iraq war when the Iraqi army attacked the Iranian oil fields in the Gulf, the price of water soared in the Arab principalities, reaching $70 a barrel in Qatar and was brought back down to $30 only after government intervention (Charnock 1984: 27). During the Gulf War, when the coalition forces initiated their military assault against Iraq, among their primary targets were Iraq's hydraulic installations, including dams, hydropower stations, and water conveyance systems, and Baghdad's modern water supply and sanitation system was intentionally ravaged. In response, the retreating Iraqis destroyed most of Kuwait's extensive oil and desalination installations and pumped enormous amounts of oil into the sea to contaminate desalination plants which provided water for the coalition forces and the population centres in the peninsula. This incident forced the Arabs in some areas to shut down their desalination plants for a while (Lacayo 1991: 32; Lemonick 1991: 40; Warner 1991: 7; Gleick 1994a: 15). After the war, however, efforts intensified to rehabilitate and expand the desalination infrastructure (Temperley 1992; WEI 1992). This is not surprising, because in the peninsula the diminishing freshwater resources inevitably mean increasing dependence on non-conventional sources of water supply and, compared to the limitations and drawbacks of other options, large-scale desalination seems to be the most practical choice. Now let us have a look at other options.

Wastewater reclamation

As a substitute for fresh water, treated wastewater has an important role to play in management of water resources in the peninsula's countries. Indeed, water from this source is already in use in many countries of the region. The widening gap between water demands and available water resources has led these countries to consider wastewater reclamation as an integral part of their water management policies. Thus, within the past few decades there has been an increased interest in the reuse of wastewater for agricultural, municipal, industrial and groundwater recharge purposes. In Kuwait, for instance, there were plans, before the Gulf War, to irrigate up to 16 000 hectares in this way. As Allan (1994a: 80) points out, these countries have the capacity to incorporate wastewater treatment systems in their water management practices, partly because they have resources to implement the expensive engineering but also because the volumes of water to be treated are relatively modest. In a comprehensive case-study of wastewater reuse in the Middle East, Banks (1991) argues that the practice of wastewater reuse in arid areas of the Middle East is now successfully established and, as a result, significant improvements in the local urban environment have been made. Similarly, Khordagui (1996: 13) and Al-Zubari (1996: 2) declare that from the early 1980s, there has been a significant move to establish the health risks involved in recycling treated effluent in landscaping and irrigation and a great effort is being made to expand these facilities to bring more land under cultivation. The emphasis on reuse of treated wastewater in irrigation is justified on economic, social, and environmental grounds. Khordagui argues that domestic sewage provides a 'convenient' and 'economic' source of water for irrigation. Examining the extent of wastewater irrigation in the region, Arar (1991), like Banks, concludes that wastewater

Table 6.5 Wastewater treatment and reuse in the oil-rich countries of the Arabian Peninsula

	Treated wastewater (mcm/yr)	Reused water (mcm/yr)	% or reused water	Total capacity (mcm/yr)	Type of utilisation
Bahrain	56.21	9.5–11	16–20	57.7	Irrigating fodder crops, gardens, and highways landscaping
Kuwait	75.92	47.23	62	76–129	Irrigating crops, landscaping highways, coastal zones, Kuwait zoo
Oman	7.5	3.96–6.33	54–86	10.6	Landscaping, parks, recreational activities
Qatar	27.5–29.5	25.2	92–86	30	Fodder crops, gardens, and landscaping
Saudi Arabia	449	100	22	>449	Irrigation of crops, highways, landscaping, and artificial recharge of aquifers
UAE	103	62	61	107.7	Irrigating parks, golf courses, highways, and urban ornamentals
Total ~	720	250	35	730–784	

Source: Al-Zubari (1996: 11)

reuse in agriculture is an accomplished and accepted fact, with a high degree of social and political consensus. Banks (1991) himself refers to different uses found for wastewater in domestic and industrial domains, arguing that the use for amenity and agricultural irrigation is in line with 'traditional features' of the pastoral activities of these societies. It is also noted that wastewater reuse could contribute to combating desertification by irrigating green belts, roadside trees, public greenery in parks, landscaping, and forestry (see Khordagui 1996: 13).

However, the figures in Table 6.5 indicate that water recycling in the Arabian Peninsula is still at its early stages. Currently the six countries of the [Persian] Gulf Co-operation Council ([P]GCC) recycle no more than 35 per cent of their total treated wastewater. Compared to their total water budget, recycled water contribution ranges between 1.4 per cent in Saudi Arabia and 21 per cent in Kuwait. Its overall average for the peninsula is slightly more than 2 per cent of the total water supply or 14 per cent of domestic and industrial demand (Al-Zubari 1996: 3, Khordagui 1996: 5). Moreover, as Table 6.5 shows, the existing wastewater treatment facilities in these countries are operating very close to their ultimate capacity: the operational capacity is near 730 mcm/yr with total treated wastewater of about 720 mcm/year.

Despite current inefficient use of treated wastewater, all the regional countries have ambitious plans for expanding the reclamation of wastewater as a significant alternative source to meet their future demands. For example, Bahrain is planning to utilise 42 mcm/year of its tertiary treated wastewater by the year 2005; Kuwait is planning to use about 140 mcm/year by the year 2010; and Saudi Arabia is to utilise about 254 mcm/year of these waters in its eastern region (Al-Zubari 1996: 3-4).

However, the expansion of this method of water management depends on three major factors. First, given the health and environmental hazards associated with the improper treatment or reuse of wastewater, further research needs to be conducted into its long-term public health and environmental implications. Second, coordination among the different agencies concerned with water supply, wastewater treatment, irrigation, public health and the environment, is a key element for further wastewater reuse. Third, public awareness and acceptance of the method are a key factor. While in many parts of the world, the concept of turning sewer water into drinking water is accepted, a study conducted in Bahrain by Madani *et al.* (1992) to assess public awareness and public attitude toward various specific uses of reclaimed water revealed that several aspects of wastewater reuse were unknown to many people, and most people were strongly opposed to using this water regardless of its quality and were willing to pay more to avoid using it. Al-Zubari (1996: 5) claims that this also is the case in other countries of the peninsula, where public awareness concerning various aspects of wastewater is generally limited and results in a negative public attitude toward some uses of reclaimed wastewater.

Importing water

A further solution among the non-conventional alternatives which are being considered to reduce the gap between water supply and demand, is the importation of water from outside the region. Importing water by tanker has already been implemented to a limited extent, and countries such as France, Japan, and the UK, which have well-placed areas of water surplus and well-established shipping infrastructure, have been trying to develop this idea. However, major problems of logistics and strategic vulnerability, along with high costs, make this option less favourable (Starr and Stoll 1988: 5). There have also been suggestions about transfer of water by pipeline from the Nile in Sudan to Saudi Arabia, from the Euphrates-Tigris in Iraq to Kuwait, from Pakistan to the UAE, from Iran to Qatar, and from the Ceyhan and Seyhan rivers in Turkey to the peninsula. Apart from the logistical, strategic and financial issues raised by such schemes, however, there are considerable international implications in each case which require the full participation and cooperation of all involved parties. For instance, agreements will be necessary on joint safeguards, on proportioning of costs and benefits, on rights of transit, and on the management of the environmental impacts of such schemes (de Jong 1989: 507).

Among the foregoing proposals, only the transfer of water from Iran to Qatar has been seriously considered. According to Bulloch and Darwish (1993: 151), this project is more economically practical and politically sensible than the others, mainly because the distance involved is much less and cooperation and coexistence with Iran is in the interest of Qatar and the other littoral countries of the Gulf. Since 1992, Iran and Qatar have been working on the project to carry fresh water from the Iranian port of Bushehr to Qatar (WE 1992: 6). The main parts of the project consist of a pipeline which takes water from the Karun river in the southwest of Iran to the coast and a 200 km underwater pipe which carries the water across the Gulf to Qatar (see Map 6.3).

Ankara, meanwhile, has underscored its role as water broker in the region by proposing to build what it calls a 'Peace Pipeline' to convey potable water from the Seyhan and Ceyhan rivers in South-East Turkey to water-poor nations of the Middle East. Brown Root, the US engineering consultancy, has prepared a feasibility study in which the cost of the project is estimated at up to $21 billion (Gowers and Walker 1989: 7). As Map 6.3 and Table 6.6 illustrate, the scheme envisages two pipelines: the western line which would deliver water to cities and towns in Syria, Jordan, and Saudi Arabia, and the eastern line which would take water to Saudi Arabia, Kuwait, Bahrain, Qatar, the United Arab Emirates, and Oman. Turkey has been promoting this idea since 1986 and during her visit to Washington in October 1993, the Turkish prime minister, Tansu Ciller, tried to revive it again.

196 *Water Politics in the Middle East*

Map 6.3 The 'Peace Pipeline' and the Iran–Qatar water project

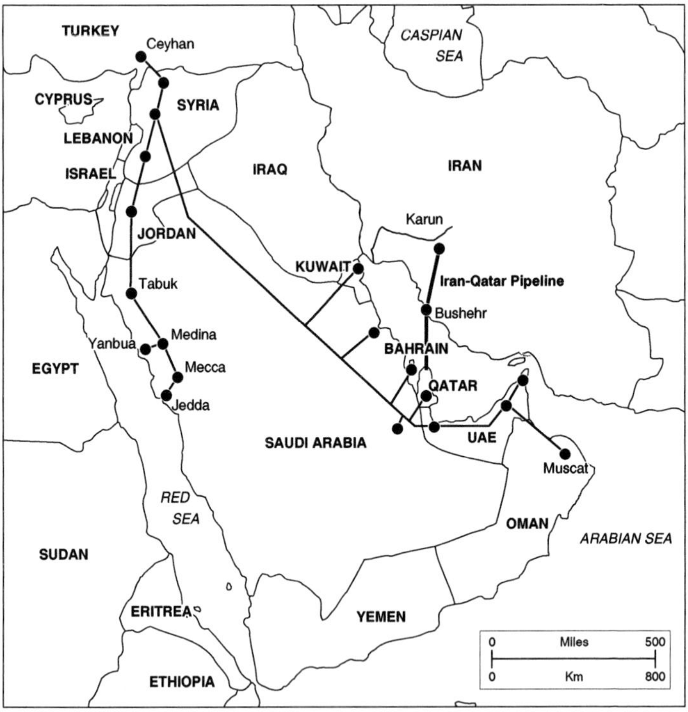

However, to be viable, such schemes cannot be decided upon unilaterally or even bilaterally. They must gain the consensus of all affected parties and assure the benefit of all involved. In the case of the Turkish 'Peace Pipeline', besides its $21 billion estimated costs, the feasibility of the project would depend on all parties reaching a broad water-sharing agreement. The greatest obstacle to such an agreement is that the downstream Arab nations do not want to place their water security at the mercy of Turkey or to rely on a life-line that would be vulnerable to attack in so many countries (Postel 1992: 82, E. Anderson 1991a: 14). Privately unwilling to rely on Turkey for drinking water, the Arabs officially argue that it is cheaper to desalinate seawater which is endlessly at their disposal. Khordagui (1996: 15) suggests that 'transporting water across international boundaries cannot be politically or economically justified'. Regarding the Turkish proposal, he argues that:

Table 6.6 The 'Peace Pipeline' capacity and delivery points

Western pipeline		Eastern pipeline	
Receiving points	Quantity (m³/day)	Receiving points	Quantity (m³/day)
Turkey	300 000	**Saudi Arabia**	
		Jubail	200 000
Syria		Ad-Damman	200 000
Aleppo	300 000	Al-Khobar	200 000
Hama	100 000	Al-Hofuf	200 000
Homs	100 000		
Damascus	600 000	**Kuwait**	600 000
Jordan		**Bahrain**	
Amman	600 000	Al-Manama	200 000
Saudi Arabia		**Qatar**	
Tabuk	100 000	Doha	100 000
Medina	300 000		
Yanbu	100 000	**UAE**	
Mecca	500 000	Abu Dhabi	280 000
Jeddah	500 000	Dubai	160 000
		Sharjah/Ajman	120 000
Total	3 500 000	Ras Al-Khaimah/	120 000
		Fujairah/Umm	
		Al-Qaiuain	40 000
		Oman	200 000
		Muscat	2 500 000
		Total	

Source: Brown and Root International, Inc., Pre-feasibility studies quoted in Musallam (1990: 16–17)

The Peace Pipeline proposed by Turkey to deliver 2.5 mcm of water to many cities in the Arabian Peninsula at a capital cost estimated at $20 billion is considered by some experts in the field as an illusion. At the same cost (using the 1993 prices), it is possible to build a number of desalination plants and the electric power plants required for their operation to produce potable water at a capacity of 7.5 mcm/day. (Khordagui 1996: 15)

Taking into account the political problems associated with the 'Peace Pipeline' scheme, it must be considered a distant prospect at best (Gowers and Walker 1989: 7). The same concern about long-term reliance on a foreign source for such a vital commodity as water, applies to other proposals regarding the possible development of international water pipelines within the region, and also to regular importation of water from outside the region by tankers or other techniques such as towing icebergs from the Antarctic. As for weather modification as a non-conventional source of water in the Arabian Peninsula, the problem is that the region basically suffers from scarcity of needed clouds.

Given the merits and drawbacks of these alternatives to enhance their fresh water resources, desalination remains the favourite choice for the nations of the Arabian Peninsula. Desalination not only provides a sense of water security but also makes the concerned countries look at the issue of water scarcity as a national and economic problem rather than a security and international issue arising out of tension in relation with their neighbours. However, as long as desalination is based on oil and gas resources, continued expansion and reliance on this non-conventional method as a supply-side water-management option is not a sustainable policy in the long term, because for both capital investment and operation of the desalination systems, the countries of the peninsula are exclusively dependent on their diminishing wealth of petroleum. As this wealth evaporates and the existing desalination plants reach the end of their operational lives, the unsustainability of this policy will manifest itself more clearly. While the fortune of the oil economies swings with the fluctuation of prices in the international markets, the [P]GCC countries need to invest at least $30 billion in the construction of new desalination plants and support facilities (Khordagui 1996: 8). Unless a major breakthrough is achieved in using solar energy or some other sustainable alternative power source for desalination, if present consumption patterns continue, these countries will face severe water shortages much earlier than anticipated.

The dawn of a new paradigm

Allan (1994a: 98) claims that the principles of demand management are nowhere evident in the Arabian Peninsula, since expensive water is still being used for the irrigation of crops that command a lower price than the delivery cost of the water applied. However, according to Ahmad (1996: 1), having realised that water will be more scarce and expensive in the future, the countries in the region are shifting their policies towards demand-side and user-focused approaches that influence consumption patterns, in order to reallocate existing supplies, encourage more efficient use, and promote more equitable access. Acknowledging that, as yet, very few countries have been able to formulate a sound national water policy, Fahim (1996: 3) asserts that there are encouraging signs indicating that the countries of the peninsula have become more aware of water problems and environmental issues. Indeed, a few decades after the disruption of the traditional institutional arrangements, a rational and environmentally sensitive paradigm is emerging in the region, albeit very slowly.

For instance, Hilary Gosh of Reuters reported, in 28 July 1997, the beginning of an intensive campaign launched to change direction in Saudi Arabia. The nation-wide campaign which is sponsored by the Saudi government aims 'to get the 18 million people living in the desert state to cut profligate water use'. King Fahd, who is himself behind the drive, urged citizens to save water as 'a religious as well as a national and development duty' and said his government was revising laws regulating consumption (see Gosh 1997). The message is clear: 'Please do not waste water'. But domestic users alone are not to blame for excessive water use, as they consume only 10 per cent of the total annual demand of around 18 billion m^3. While the report indicates that per capita water consumption in Saudi Arabia is higher than in other countries at the same stage of development, Khordagui (1996: 17) argues that 'the region already has the lowest per capita consumption rates in any arid area of the world'. He believes that 'policy makers looking to increase political flexibility by reducing demand would be hard pressed to find meaningful cuts in the domestic sector'. The truth is that farmers are the main culprits.

The kingdom's five-year development plan to the year 2000 indicates that 'agriculture continues to consume almost 90 per cent of water from all sources, and the expansion in grain production has been heavily dependent on non-renewable groundwater in most areas'. The plan calls for a decline in the use of non-renewable groundwater to 13 billion m^3 from about 14.8, and a cut in water use from 18.2 billion in 1994 to 17.5 billion m^3 by the end of the century. These targets are to be achieved by reducing the rate of water consumption in agriculture at an average annual rate of 2.2 per cent and achieving balanced consumption rates for other purposes, the report indicates.

It is now admitted that since the 1980s, Saudi Arabia has been wrongly subsidising production of many crops not suitable to its arid environment. For example, in 1992 the government spent $2 billion in subsidies for the domestic production of 4 million tonnes of wheat, offering farmers a support price five times the international price. However, the government is beginning to realise the harmful environmental consequences of this policy and is gradually reducing subsidies on agricultural production (Ahmad 1996: 4). According to the Reuters report, as a result of new policies, production of wheat decreased to 1.3 million in 1996–7 and the government is planning to reduce production of other water-intensive crops such as forage and alfalfa. Other demand management techniques such as adjusting the water tariffs so that people can understand the real value of water, and imposing penalties on abusers are contemplated. To sum up, an evolving awareness of the economic and environmental costs of water and of the need to derive sound and efficient economic returns from scarce water used in national economies is emerging in the peninsula. The concerned countries are coming to terms with the fact that they need to set a conceptual and institutional framework for addressing water issues.

Conclusion

In this chapter, we claim that in extremely arid climatic conditions such as the Arabian Peninsula, where irregular scanty precipitation creates an impossible situation for a perennial surface water system to exist, water scarcity becomes a catalyst for cooperation among the concerned parties to safeguard their precious water resources. The historical background of water management in the Arabian Peninsula indicates that, faced with problems approaching critical severity, communities devise solutions or expedients to fit. This hypothesis has been tested and proven in this hyper-arid region time and again. Of course, rapid socio-economic and technological developments may drastically change the traditional lifestyle and the demographic configuration in a society, as it did in the Arabian Peninsula after the discovery of petroleum. And these dramatic changes can disrupt the ability of the traditional norms and institutions to manage the ever-increasing water demands. However, the recent experience in the peninsula suggests that even in such circumstances, the shared need for optimum management of this vital resource can become a source of accord and integration rather than rupture and conflict. For instance, due to the alarming depletion of its groundwater resources, in summer 1983 Bahrain faced a critical water crisis. This incident not only did not provoke a dispute between Bahrain and Saudi Arabia who share the same aquifers with Bahrain, but resulted in the Saudis helping Bahrain to expand its desalination capacity and avert the crisis. Moreover, since the mid 1980s, the [P]GCC has been studying the problems of both water and electricity supplies in the Gulf states, and attempting to work out a Gulf-wide strategy which might include a [P]GCC grid system for both utilities. Given that all the member states face a broadly similar problem over water resources, observers are convinced of the need for such a Gulf-wide water strategy (see Bodgener 1984: 38).

Furthermore, though there have been some minor border disputes in the peninsula, no documented case of violent conflict or open dispute over water has occurred in the region. This is in spite of the fact that the countries of the peninsula have the poorest per capita water availability and the highest per capita median cost of water supply and sanitation in the world, and also lack any binding international water agreement which could regulate the appropriation of shared water resources among them. This situation clearly casts doubt on the theory of 'water war' and the predictions that the Arabian Peninsula is an area where conflict over water is likely to arise (Starr and Stoll 1988: 6–7; E. Anderson 1991a: 11; Bulloch and Darwish 1993: 24). However, this is not to deny that when there is serious conflict in the region, water may become a strategic commodity to be denied to an enemy.

As noted above, this strategic importance of water in wartime conditions is not a new phenomenon. But the catastrophic environmental consequences of the Gulf War, particularly the deliberate destruction of water supply systems, led observers such as Gleick to argue that:

> The Persian Gulf War underscored the many connections between water and conflict. During this war, water and water supply systems were targets of attack, shared water supplies were used as instruments of politics, and water was considered a potential tool of warfare. The dams, desalination plants, and water conveyance systems of both sides were targeted for destruction. Most of Kuwait's extensive desalination capacity was destroyed by the retreating Iraqis. Oil spilled into the gulf threatened to contaminate desalination plants throughout the region. And the intentional destruction of Baghdad's modern water supply and sanitation system was so complete that the Iraqis are still suffering severe problems as they rebuild them. (Gleick 1994a: 15)

However, none of Gleick's points suggest that water conflict was the *cause* of the Gulf War: they only indicate that water was a military target during the fighting. Moreover, as Dellapenna points out, such targeting of water supplies was only carried out because reciprocal Iraqi retaliation was impossible:

> Each side even in intense conflict situations realises that depriving the other side of the available water necessary for survival is one of the few steps that could make even a significantly weaker state desperate enough to fight against any odds and to target its enemies' water facilities, facilities that would be impossible to defend... against a sufficiently determined foe. Only in the recent Gulf War did one side (the coalition) target the water facilities of the other side (Iraq), in large part precisely because the militarily dominant partner in the coalition did not fear reciprocal attacks on its own facilities. (1995: 56)

The war also demonstrates that the hypothesis suggested by Bulloch and Darwish (1993: 161) that, 'whoever controls water or its distribution can dominate the Middle East and all its riches' is still far from being true. There is no doubt that, at least in this part of the Middle East, the 'black gold' – oil – is still much more desirable and commanding than the 'white gold' – water. So while oil has proved to be capable of provoking regional countries as well as the international community to ignite the fire of war, water has never been a catalyst for war in this region.

Nevertheless, as Bulloch and Darwish (1993: 195) declare, the lessons of the Gulf War – in particular the vulnerability of desalination plants – are now being absorbed and acted upon. One of the main precautions that has been seriously contemplated is to secure a strategic water reserve, in case desalination ever has to be stopped. Other measures concerning demand management and water consumption modification are also under review. However, the success of recent changes in water policies in the Arabian Peninsula and other international water basins of the Middle East is dependent upon wider issues of development, government, and governance in these societies. As Fahim (1996: 5) points out, these are sociological issues that seriously affect water policies in the region and, therefore, should be addressed sociologically. Although this task is beyond the scope of this study, in the concluding chapter we shall briefly touch on some of these issues.

Notes

1 For detailed information on the hydrology of the Arabian Peninsula see: Al-Alawi and Abdulrazzak 1994; AOAD 1996; ECSCWA 1994; ECWA 1978; FAO 1979; Gischler 1979; Burdon 1982; and Shahin 1989).
2 The 'wadi' is the Arabic name for a valley or stream bed that remains dry except during the rainy season. It also refers to the stream that flows through such a channel.
3 See, for instance, the divergent figures presented in the reports and studies cited in n. 1.
4 Petroleum production in the Middle East began in May 1908 in Iran. As for the Arabian Peninsula, however, it took more than a quarter of a century to discover its underground treasure. Oil was first exploited in Bahrain (1932) and followed by Saudi Arabia (1938), Kuwait (1938), Qatar (1949), Oman (1954), UAE (1958), and recently in Yemen (1987). (For a comprehensive account, see Yergin 1991.)
5 There is a distinction between Yemen, Saudi Arabia, and Oman which traditionally have had some significant agricultural resources, and the smaller sheikhdoms of the peninsula where the water constraint is much more immediately binding and the role of farming as an economic activity was marginal even before the oil era.
6 It should be noted that there is always a risk of double entry in collection and processing of hydrological data. In this case, if we consider that surface water is one of the major contributors to groundwater recharge, the real water deficit may be much more striking than it appears to be. However, a significant portion of water used in irrigation or in the domestic sector returns to underground aquifers. This could be accounted for in terms of quantity but, since the returned water is always contaminated, it degrades the existing freshwater resources. Not surprisingly, therefore there is often a wide range of discrepancies between different sources of hydrological data in the Arabian Peninsula.
7 Although treaty provisions and international law regarding groundwater resources lag far behind the progress of international law on surface waters, some progress has been made in this area with the Bellagio Draft Treaty of 1989. With the goal of advancing international law and institutions, a multidisciplinary group of specialists, over an eight-year period, drafted an international groundwater convention. The draft, entitled *Transboundary Groundwaters: the Bellagio Draft Treaty* consists of 20 articles which cover issues such as quality protection, depletion, drought, transboundary transfers, withdrawal and recharge, and existing rights and obligations. The fundamental goal of this draft treaty is 'to achieve joint, optimum utilisation and avoidance or resolution of disputes over shared groundwaters in a time of ever-increasing pressures upon this priceless resource' (Hayton and Utton 1989: 663). The draft suggests that uncontrolled and unilateral withdrawal from international aquifers should be abandoned; rights to such resources and their rational consumption and protection should be determined by mutual agreement; and international aquifers in critical areas should be managed by joint resource management machinery in order to avoid competitive extraction.

8 According to the World Bank, the Middle East has the highest median cost of water supply and sanitation in the world: the costs reached a median of $300 per capita in 1985, about twice those in the US and more than five times the costs in southeast Asia (Hillel 194: 36; Starr 1995: 197). This is largely because of the enormous capital outlays for desalination plants and their maintenance expenditure.

9 The Saudis have been criticised as the most 'egregious' water user in the world today (see Postel 1992: 31). The water quantities they are pumping for irrigation are much more han 14 bcm/yr of fossil water. Such an excessive mining of groundwater is encouraged by the government to boost domestic food production. By heavily subsidising land, equipment, and irrigation water, and by buying crops at several times the international market price, the government has sought to achieve food self-sufficiency. For instance, in 1992, offering farmers a support price five times the world price, the government spent $2 billion of subsidies on production of 4 million tonnes of wheat, which was twice the domestic requirement (Ahmad 1996: 4). As a result of this policy, the Saudis, who until the mid-1980s were the third largest food importer, in 1991 reached the rank of the world's sixth largest wheat exporter (Beschorner 1992: 72). Though the kingdom still imports barley and other food crops, its drive for food self-sufficiency is the single biggest drain on the peninsula's water resources. For more insight into the agricultural policy of Saudi Arabia see Al-Tawati (1990); Duwais (1990); E. G. H. Joffe (1985); Joma (1991); *Middle East Economic Digest* (1984: 24–5; 1985: 78ff; 1988a: 22; 1988b: 19ff); Temperley (1992). For a detailed review of plans and strategies in other countries of the peninsula see Al-Ruwaih (1995); Al-Kuwari (1996); Al-sheikh (1996); Bodgener (1984); Drake (1988); Lichtenhaler (1996); McLachlan and Beaumont (1985); Saif (1987); and Christopher Ward (1997).

10 Although the underground water resources of the peninsula have been subject to numerous studies, their storage capacities and rates of recharge and transmissibility are not yet determined. According to Khordagui (1996: 4), deep groundwater reserves are estimated at 2175 bcm with the major portion (1919 bcm) located in Saudi Arabia. He also holds that recharge for all the aquifers of this type is estimated at 2.7 bcm/year. However, Temperley (1992: 19) estimates that there is near 2000 bcm of water in these aquifers some of which take hundreds, even thousands of years to replenish.

11 For a brief analysis of the potentials and constraints associated with each non-conventional water resources development option, see Khordagui (1996: 9–14).

12 Although as early as 1938, Saudi Arabia bought two condensers for desalting the drinking water supply of Jidda, its largest city and principal port, Kuwait was the first country in the region to embark upon mass production of desalinated water as its major source of drinking water. The first unit of the multistage flash distillation was commissioned by Kuwait in 1958 and Saudi Arabia joined the club in the late 1960s (Khordagui 1996: 4; Temperley 1992: 22). It should be noted that fresh water can be derived from salty sources by five methods: multistage

flash distillation, multi-effect distillation, reverse osmosis, vapour compression distillation, and electrodialysis. Of these, the first three are the most common. For a detailed analytic and comparative study of desalination see Dabbagh *et al.* 1994: 203–42.

13 There are some discrepancies between data presented by western and local observers on these figures. For instance, Postel (1992: 45) and Starr (1995: 54) believe that 60 per cent of the world's desalination capacity is concentrated in the Gulf, with Saudi Arabia's capacity alone exceeding 30 per cent of the global production. But Al-Zubari (1996: 3) claims that 'at present the total capacity of desalination plants in the GCC countries is about 1852 mcm/year' – equivalent to 33 per cent of the world's capacity. while Khordagui (1996: 5) declares that GCC countries house almost half (49.5 per cent) of the world's total capacity, with Saudi Arabia alone contributing about half of the GCC desalination capacities (25 per cent of the world capacity).

14 The cost of desalination depends on several factors: the method of desalination, the capacity of the plant, and the salinity of the input water. The more energy-efficient the method, the larger the capacity of the plant, and the less salty the input water, the cheaper the process of desalination is. Advantages, limitations and costs of desalination process are discussed in Dabbagh *et al.* (1994); Temperley (1992); and Weber (1991).

7
Conclusion

Summary

While in the humid and temperate parts of the world, water was regarded until very recently as a free gift of nature which was always available in abundance, people in the Middle East since antiquity have appreciated the socio-economic value of water, and have treated it as a valuable finite resource to be revered as the essence of life. Indeed, in this part of the world, water has been the fundamental factor in determining the rise and fall of civilisations, and its scarcity, as an environmental phenomenon, has deeply affected their social, cultural, economic, and political structure (see Chapter 1). However, the decisive role of water is no longer confined to arid and semi-arid regions. During the last few decades, due to expansion of populations, agro-industrial pollution and aspirations for greater economic development, in other parts of the world the pressure on limited fresh water supplies has approached the 'water barrier' beyond which the need for safe water becomes a dominant concern. Indeed, one can hardly find a country which is not, in one way or another, at the edge, if not in the middle, of a water crisis. This phenomenon has severely affected the health and very survival of many people[1] and deeply influenced the course of social, economic, and political development in many countries including some relatively water-rich countries (see Chapter 3). Furthermore, climatic changes and the harmful ramifications of global warming may well be exacerbating this predicament.

Apprehension of this global trend and the increasing realisation of the importance of water in the continued well-being and development of humankind and its effects on international relations, has turned the issue of water resources management into a high profile international political issue. Indeed, since March 1977 when the United Nations Water Conference was held in Mar del Plata, Argentina, the international community has been engaged in numerous activities at international, regional, and local levels to promote the preparedness of nation-states to avoid water crises or to alleviate their consequences (for a brief inventory of these activities see Bailey 1996: 2–18). Not surprisingly, therefore, during the last two decades, the water crisis, which is a major subset of worldwide environmental problems, has been the focus of innumerable studies and investigations from a variety of perspectives.

Focusing on five distinctive approaches in the literature of hydropolitics – namely security, economic, legal, technological, and environmental – Chapter 2 shows that most observers believe that water is becoming one of the universal problems facing humanity into the twenty-first century. Those who approach the issue from a national security perspective have repeatedly warned that *wars* of the next century will be *over water* and they have predicted that the Middle East will be the focal point of this grave international crisis. Indeed, many observers have suggested that the water conflict is deeply rooted in the history of the Middle East (see, for instance, Nijim 1990; Bulloch and Darwish 1993; Hatami and Gleick 1993; Starr 1993) and several of them have securitised and militarised the issue to the point of making war seem inevitable (see Thompson 1978; Cooley 1984; Falkenmark 1986a; Musallam 1990; Gruen 1991; Starr 1991; Bulloch 1993; Lowi 1993a; Gleick 1994a; Pope 1995).

However, if we look more closely at the alleged link between water scarcity and international conflict, it is evident that water scarcity is neither a sufficient nor a necessary cause of violent conflict, since violence may occur in situations of water abundance and it may not occur in situations of water scarcity. Indeed, as we noted in Chapter 2, some water-rich countries such as those in Indochina are racked by conflict, and some water-poor countries such as the Netherlands are very peaceful, while both are located in international river basins (Lipschutz 1992: 12–13). So, if water scarcity contributes to conflict, it always operates within a context that is formed by other determining political, legal, economic, and cultural factors; and any analysis that disregards the context and attempts to suggest a universal explanation for water conflict is doomed to failure. As Lipschutz and Holdren (1990: 126) point out, to suggest that resource scarcity is 'a direct cause of international conflict is to ignore most of the other forces and motives that impel decision-makers to action'.

If we look at the conflicts of these 'other forces and motives' we find that, as Chapter 2 indicates, there are less gloomy approaches to the issue of water crisis that identify various mechanisms for averting the crisis and resolving the conflicts. For example, the economic approach suggests that the water crisis is avoidable if we treat water as a valuable commodity, like oil, not a free resource like air. However, the insufficiencies of market forces, particularly in the context of sustainable development, in dealing with such a vital resource make this approach problematic. The legal approach addresses the issue of water rights as the root cause of the water crisis arguing that, due to lack of treaties or agreed-upon international legal structures for settling disputes over shared resources, conflicts of interests turn into confrontations in the context of regional and global power balances. However, this approach ignores the fact that even after translating legal theory into practice, the social and cultural norms and values continue to affect the way people consume water. The technological approach seeks the solution in increasing the efficiency of the distribution and supply systems. However, as long as this approach ignores the context of sustainable development, it is merely addressing the symptoms rather than the causes of the crisis. This brings us finally to the fifth approach – environmentalism. The environmental approach marks a fundamental shift from the mind set of the other four approaches, in that its perspective is holistic, not partial, and it entails the adoption of a 'water ethic' within the framework of the principle of sustainable development. As such, the environmental approach alone gets to the root of the water issue – human behaviour in relation to nature.

Two hypotheses

The following exposition recapitulates and orders some of the main arguments developed during the course of this study and establishes two hypotheses: one related to the past and the other related to the future.

The past

Although the most frequently cited hypothesis in the literature of hydropolitics is that water scarcity has been the cause of conflict and war in the Middle East, *we argue that water scarcity has virtually never caused armed conflict in the region.* Indeed, although water has sometimes provoked tension and dispute in the Middle East, it has much more often promoted co-existence and cooperation between indigenous communities, and *we argue that water scarcity has invariably been a platform for cooperation in the region, through which all riparian parties have achieved greater gains.* As a win–win strategy, therefore, the cooperative sustainable development of the region's common water resources has reinforced regional integration and peace.

The future

Recent developments in the Middle East indicate that a new environmental paradigm is emerging in the region and, as the countries in the region come closer to the end of the supply-side management phase and turn to demand-side measures for managing their water resources, efforts are being made to draw attention of users to the value of water and to encourage them to allocate water to activities that achieve the best returns. Moreover, from this new perspective a holistic understanding of resource problems, including their security implications, in terms of environmental security which surpasses the traditional notion of national security, will empower countries to manage their environmental problems through accommodation and without recourse to conflict. *We argue, therefore, that no matter how acute the water crisis becomes in the Middle East, the new environmental paradigm will ensure that it is resolved without resort to armed conflict.*

Let us look more closely at these two hypotheses.

Hypothesis 1: not conflict, but cooperation, in the past

Many important points and arguments raised in the preceding chapters serve to substantiate the first hypothesis. A systematic analysis of the norms, regulations, and civil institutions established in the Middle East to deal with water resources management showed that although uneven precipitation, recurrent severe floods and droughts and the extreme fluctuations of the rivers and water tables have always been the main concern of the region's inhabitants, the civilisations that flourished in this region successfully came to terms with their natural environment. We clearly demonstrated that the traditional Middle Eastern civilisations had not only the engineering skills necessary for an efficient water management system, but also the social prowess and legal institutions required for maintaining the functionality of such system. These analyses establish that contrary to what is conventionally believed, *water resources have never been a source of strategic rivalry or the root cause of military conflict in this region.* As Wolf (1997: 1) points out, this conclusion applies across the globe; 'only seven skirmishes are found throughout the world; no war has ever been fought over water'. Indeed, the historical evidence suggests that, having realised the life-sustaining role of water, the indigenous inhabitants of such an arid environment deliberately set out to cooperate and coordinate their collective efforts in a systematic way to control the hydrological chaos of the region for the sake of all riparian parties. Cisterns in the Jordan Basin, *qanat* systems in the Arabian Peninsula, and hydraulic projects in Mesopotamia are symbols of these cooperative efforts. Of course, it is true that in times of war and hostility precipitated by factors *other than water-related issues,* water has been used both as a destructive weapon and as a defensive barrier. However, the desire of kingdoms and empires for expansion of their territories or their passion for subjugation of other states is very different from their going to war in order to capture scarce water resources in a specific area. There is a sharp difference between conflict over a shared water body, and using water as a military target or as a weapon against an enemy.

The assertion that the Middle East is the region in which conflict over water is at its most extreme[2] is based on speculation rather than on hard evidence. A typical example of such speculative arguments is provided by Gleick (1994a: 8), who claims that 'water can become a source of strategic rivalry because of its scarcity, the extent to which the supply is shared by more than one region or state, the relative power of the basin states, and the ease of access to alternative freshwater sources'; that 'in the Middle East, water is scarce and widely shared by countries with enormous economic, military, and political differences... [and] few economically or politically acceptable alternative sources of supply'; and that 'the temptation to use water for political or military purposes has often proved irresistible'. Many other writers have similarly tried to demonstrate that there is a high incidence of water wars in the Middle East. As pointed out in previous chapters, however, the assumptions of this argument are questionable. What we want to add here is that this type of argument reflects a secular industrial Western cultural view which ignores the sociocultural values and norms in other societies. Let us explain this point.

If water shortage itself was a sufficient provocation to bring two countries to the brink of war, there should have been several historical cases of water war in the Arabian Peninsula since it is one of the most *parched* areas of the world, where scarce water resources are *shared* between seven countries with considerable *disparity* between their economic, political, and military capabilities. Similarly, in North Africa which has been always the victim of long-lasting and recurrent droughts, we should have hard evidence of such so-called water wars. But in fact, there have not been water wars in either region. This indicates that, although, as a general rule, scarcity of resources may intensify competition between states, it does not *necessarily* lead to *violent conflict*. Rather, the outcome of competition very much depends on the sociocultural and political context in which the parties compete. For instance, communalism has been an important mode of thinking and of managing resources in many societies throughout the world, from the nomads of the Arabian Peninsula to native American people. This is why traditional resource management systems are often community-based (Berkes 1989: 5).

The Islamic culture which is the dominant culture in the Middle East emphasises responsibility to the community, rather than the unbridled individualism prevalent in some Western industrial cultures. The Western culture is a context in which the emphasis is upon individual identity, endowed with opportunities, and the challenge to compete within a totally secular market economy. By contrast, in Middle Eastern societies, the primacy of a communal identity which is based on family, clan, tribe, nation, or religion, has long been established. Hence it may not be surprising for American society to witness armed farmers sitting atop the Los Angeles water gates, blowing up the city's aqueduct 17 times to prevent drainage of water from their lands, and engaged in confrontation with state militia in a water conflict that continues for decades (Trager 1994). But in the Middle East, where hydrological realities have deprived most of the societies of sufficient exclusive water resources, and where water is regarded as a gift of God which cannot be monopolised by anyone, water scarcity encourages cooperation, not confrontation, to find a solution to the common predicament. The historical evidence indicates that in these societies the sociocultural values transform water competition into a cooperative

212 *Water Politics in the Middle East*

win–win game. This is not to deny that in some circumstances conflict may occur in these societies. Rather, the point is that where in a society or in a region people can manage to establish appropriate principles and institutions to prevent water conflict or to resolve it as soon as it occurs, the possibility of such conflict will substantially decrease. The facts presented in this study suggests that in general the Middle Eastern societies have been remarkably successful in creating such institutions. This is why even the sceptics sometimes (inconsistently) acknowledge that 'water related disputes in the Middle East and elsewhere are more likely to lead to political confrontations and negotiations than to violent conflict' (Gleick 1994b: 50).

Despite the general atmosphere of political tension and the rapid socio-economic developments in the Middle East, therefore, little open conflict over water has erupted in this volatile region. This point has been so significant that some observers have called it a paradox or contradiction, acknowledging that 'complexities and tensions raised by hydrological problems have often tended to compel co-operation where other non-water antagonisms have degenerated into warfare' (Naff and Matson 1984: 4), and arguing that 'not only did co-operation between Israel and Jordan subsist in the years leading up to the recent peace treaty between the two countries, but it is perceived by geographers that the water issue would be one of the easiest to overcome if the Israeli–Palestinian conflict would reach a peaceful resolution' (Shapland 1995: 305).

Indeed, water scarcity has actually been a stimulus to international peace making.[3] Many countries that are involved in water disputes have used the issue as a lever, or instrument, by which political goals can be accomplished. External powers have also used the water issue as a suitable means by which they can have a voice in the region's affairs. Ironically, this less-recognised aspect of hydropolitics in the Middle East was exposed by one of those observers who until recently called for US intervention to mitigate the water crisis in this region. In her latest reflection on water issue in the Middle East, Joyce Starr angrily revealed how misguided US diplomacy torpedoed one of the region's most constructive and long elaborated attempts to deal with its critical water shortage: the Middle East water summit sponsored by Turkey's late President Turgut Ozal and supported by the World Bank, the United Nations Development Programme (UNDP), and the United Nations Environment Programme (UNEP), due to be convened in Istanbul in 1991. It was supposed to include Israel as well as the Arab states, two-thirds of whom agreed to attend, including Jordan and Egypt. Its goal was to create a common framework for water security and a regional dialogue on water management including discussion on 'a regional water community' similar to the European Coal and Steel Community created by the Marshall Plan after the Second World War. Among its suggested functions was coordination of water planning in the region, regional data collection, cooperative research, emergency response mechanisms, dispute resolution mechanisms, and water and environmental 'peace crops' for children in the region. This initiative promised to have far reaching effects on the peace process in the Middle East, but it was aborted because it was perceived by some US officials to endanger the 'official' peace process. If Israel were allowed to attend a multilateral dialogue with Arab states outside the formal peace framework, critical US leverage would be lost over the Israeli government (Starr 1995: 107–21).

Not only has direct intervention of foreign powers turned opportunities into obstacles, but their indirect involvement in the region has also been a cause of conflict. A comparison of the three case-studies in Chapters 4, 5 and 6 illustrates this point. Consider the countries of the Arabian Peninsula, which not only share a hyper-arid environment, but also cope with a burgeoning population and rapidly depleting oil and water resources with extremely limited options and alternatives at hand (see Chapter 6). According to the conventional realist perspective, this situation is a recipe for war. However, despite some thorny border disputes, there has been no sign of water conflict leading to severe interstate violence in this basin. This is partly because the political orientation of these countries, especially their foreign policy considerations, has been identical – adherence to Anglo-American foreign policies. Except for Yemen, this region was never a site of East–West rivalry that imposed itself on riparian parties in other water basins of the Middle East. Hence, it escaped externally induced tensions that could have generated water conflict.

The Euphrates–Tigris basin, by contrast, is endowed with relative abundance of water and still has room for further development. Hence, the lack of firm and comprehensive cooperation among the riparian countries in this basin was the outcome of a host of different variables (as discussed in Chapter 5), including the intense East–West rivalries that grafted itself on to the political resentments and animosities which already existed in the region, rather than a consequence of water scarcity. It was East–West rivalry for economic and technical supremacy in the region that gave strategic significance to the development of the Euphrates, and it was in the 'East–West' context that for 'political reasons' the contracts were signed for the construction of competing hydraulic projects in three riparian countries. However, it could be argued that the lack of overall water scarcity in this basin provided a particular circumstance in which the parties could use the water issue as a bargaining chip in their negotiations, and in due course as a political leverage, without frustrating each other too much. This would explain why downstream states in the basin do not resort to military force to prevent upstream developments as long as water is the only bone of contention. Nevertheless, given that the region was a scene for the playing out of East–West confrontation, and given the territorial disputes and local insurgencies such as the Kurdish issue in the basin, it is interesting that the importance of water has encouraged the three parties to continue their technical consultations and political negotiations since the early 1960s. Hence, despite the potentially destabilising effect of Cold War politics in the region, the centrifugal pressure of water politics has enabled peaceful negotiations to prevail. This negotiating process has since been enhanced by the end of the cold war, which has undermined the likelihood of conflict between riparian states and prepared the ground for more cooperative arrangements in the future.

The Jordan River basin differs significantly from both the Arabian Peninsula and Euphrates–Tigris basin. Unlike the Arabian Peninsula whose abundance of oil resources has enabled it to substitute energy for water through unconventional methods so that it could produce desalinated water in volumes much more than the whole discharge of the Jordan River, the Jordan basin is deprived of both oil and adequate water resources. And, unlike Mesopotamia, which has still many opportunities for further development of its water resources, the Jordan basin has already passed the red line and any further water supply has to be found in unconventional methods. Moreover, the Jordan River basin has been the locus of the most intense international conflict during the last five decades. Not surprisingly, therefore, the Jordan basin is the most frequently cited case amongst all the water systems in the Middle East whose scarce waters have been widely tipped as a catalyst for serious conflicts and the likeliest of all to create a new water war. However, as explained in Chapter 4, this is a very specific context in which water issues are part of a complex ethnocentric and territorial conflict between Arabs and Israelis – a conflict which until 1990 was itself part of a wider East – West dispute. The climatic conditions of the region, the instrumental attitude of Zionists towards the environment and their determination to alter the hydrological map of the basin so that it could accommodate all the Jews who wanted to settle in Israel – the ingathering that was the only rationale for the establishment of the Jewish state – and the Arabs' perception that their land and water resources were sources upon which the illegitimate development projects of their bitter enemy, Israel, depended, created an explosive background of belligerency in which hydropolitics merged into geopolitics and produced the theory of 'water war'.

While no one can ignore the role of water in shaping the prolonged process of the Arab–Israeli conflict and its decisive part in the current peace negotiations, a detailed analysis of the evolution of hydropolitics in the Jordan basin exposes the inadequacy of 'water war' theory. As shown in Chapter 4, since Zionism and its imposition on the land of Palestine by external powers has been the foremost and fundamental *cause* of conflict generation in this basin, the *contribution* of every other factor to protraction or intensification of the *core issue* of the Arab–Israeli conflict is of secondary significance. In circumstances where the very existence of one party is deemed illegitimate by the other party, and there is no ground for co-existence between them, any hypothesis that suggests that the threat to posterity and survival that water scarcity represents is an *intrinsic security issue* for the parties, and any suggestion that attributes a *causal* role to water and tries to quantify the *causative* contribution of water to exacerbating the hostilities and provoking wars between riparian parties, is unconvincing.[4] Indeed, the experience of riparian parties in the Jordan basin indicates that even in this tense context 'the "superordinate goal" of mutual survival may compel the hostile neighbors to cooperate' (Kolars and Mitchell 1991: xxvii).

As shown in Chapter 4, there are signs of a more 'environmental' approach to water scarcity occurring in Israel, and this bodes well for a cooperative strategy. It may even be that a solution to the water conflict will help to break the log jam in the overall peace negotiations. As Hillel (1994: 287) points out, a dispute over water resources, rather than being a justification for war, may precipitate moves towards peace by inducing non-violent resolution through compromise and collaboration. He quotes the Israeli politician, Abba Eban, as saying that 'the nations of the region will act rationally, once they've run out of all other possibilities.' The truth is that water scarcity and quality degradation have not become a source or trigger of conflict in the Jordan basin; rather their cumulative negative effects have inspired environmental awareness, and pushed the concerned countries towards cooperative arrangements to assuage the crisis (see Chapter 4). The Israel–Jordan peace treaty of 26 October, 1994 which is one of the highlights of the current Middle East peace negotiations, shows that the most water-poor parties in the region can overcome the zero-sum game on the distribution question and make it a win–win compromise. Settlement of bilateral water disputes is a centrepiece of the agreement which is addressed in the main body of the treaty, and water issues fill all of Annex II of the treaty. Article 6 of the agreement recognises the 'rightful allocations' of their common water resources and the opportunity that 'the subject of water can form the basis for the advancement of cooperation between them'. Acknowledging that their existing water resources 'are not sufficient to meet their needs', the treaty provides a framework for future cooperation in all aspects of water management and development, including development of new water resources, minimising wastage, preventing pollution, dealing with shortages, as well as data exchange and joint research. Furthermore, as Libiszewski (1995: 76) affirms, 'the treaty contains a number of concrete stipulations aimed at creating functional interdependencies between the parties' and 'when implemented, these joint water projects will have the potential to cement peace between the two countries'.

In this way, a vision of regional environmental cooperation has been encouraged as an alternative to the sterile conflicts of the past. Since all parties are exploring ways of increasing supply within serious economic and environmental constraints, 'sharing of expertise, opening access to hydrologic data, and exploring joint water conservation and supply projects offer the best opportunities for reducing the risk of future tensions over water in the Middle East' (Gleick 1994a: 35). Further evidence of this cooperative discourse over water scarcity is contained in the framework of the Middle East peace process, where, besides the bilateral negotiations between Israel and its neighbours, a multilateral track of negotiations has been designed to deal with five major issues at the regional level: management of regional water resources, the question of refugees, environmental problems, regional economic development, and regional arms control. In this framework, the working group on water

Conclusion 217

resources focuses on technical matters such as data gathering and its exchange, improvement of water supply, water management and conservation, and planning for regional projects, while the environmental group concentrates on combating marine pollution, halting desertification, safeguarding drinking water, and wastewater management (Peters 1994).

Although these multilateral talks on water have been overshadowed by political obstacles between Israel and two upstream states of the Jordan River – Syria and Lebanon – who have boycotted the negotiations, the working group has discussed various plans and proposals which could become fields of cooperation in the future. For example, Oman proposed to establish a regional research centre on desalination technologies in Muscat, and a USA/EU proposal urged the need for establishing regional hydrological data banks. The environment working group has been more successful in establishing a regional framework for cooperation; in its sixth round of talks in Bahrain in October 1994, the group was able to approve unanimously an Environmental Code of Conduct to govern the future direction of economic development and environmental programs and legislation in the region, and to serve as a basis for research and scientific development. Moreover, a Canadian suggestion for building up an environmental impact assessment infrastructure in countries of the region, and an action plan by the World Bank for collaboration to control and natural resource degradation and desertification, have been into consideration (see Libiszewski 1995: 84; Amir 1995; Warburg 1995).

Hypothesis 2: the new paradigm – environmentalism – for the future

Following on from this last point, we argue that the future is best viewed in terms of this environmental perspective. Inspired by the revitalising global environmental movement, cooperation is already in evidence in the emerging approach to environmental issues in the Middle East. Interpreting these developments from a traditional perspective of national interests or national security is increasingly seen to be neither helpful nor conclusive. Indeed, as we discussed earlier, using language traditionally associated with violence and war to understand environmental issues is now regarded as seriously misleading. This is why 'environmental security' and 'national security' are considered as opposed perspectives arising in the context of alternative world-views (Dyer 1996). Therefore, increasing public awareness about the crucial role of environmentalism in providing safe, prosperous, and stable living conditions for humankind highlights the issue of sustainable development by which the traditional modes of development are questioned. The much vaunted concept of the 'sustainable society' implies that the traditional norms and perspectives, and the old paradigmatic relationship between humans and their living environment, are seriously flawed. As Van Dijk (1994: 56) rightly argued, 'to understand the present crisis, we should become aware of the route which has led us into this crisis'.

Certainly, the days when environmental issues were interpreted as technical defects in our social machinery are now in the past. It is increasingly recognised, by both public and decision-makers, that we are confronted with a very complex problem which is closely interwoven with our entire civilisation (Boersema 1994: 20). By introducing the notion of environmental ethics – a new ethics which is essentially different from an ethics developed from within traditional theories such as 'liberalism', 'realism', and 'Marxism' – the environmental discourse challenges the most basic values of contemporary Western civilisation: the nation-state and its privileged status in world politics; the notion of nationalism; the assumptions of resources in nature as available for limitless exploitation; ever-increasing consumption and the belief that affluence is intrinsically superior to any other life-style; and faith in 'the market'. Globalisation has been one of the most important stimulants of environmentalism; from a global perspective, ecological decay and environmental crises do not so much threaten national security as challenge the utility of thinking in national terms. Therefore, 'thinking of the environment as a national security problem risks undercutting the sense of world community and common fate that may be necessary to solve the problem' (Deudney 1991: 25–6; Dalby 1992b: 114).

Interestingly, the Islamic perspective shares many of the basic insights of the environmental discourse: that the roots of our environmental crises, and therefore their solution, lie in humanity's conception of its role in the overall scheme of the creation; that these crises have reached a critical point so that humankind have to confront certain basic questions about their relationship to the environment; that these are not questions of technology, but questions about the fundamental nature of humanity, and the nature of the universe in which humankind exists; and that the present environmental imbalance represents a philosophical and ethical crisis of Western civilisation (see Dolatyar 1996). Since the traditional norms and values inherited by Middle Eastern societies are based on the Islamic ethos which emphasises communalism and universalism, it seems that many of the norms and values introduced by environmental ethics (Attfield 1983; Elliot and Gare 1983; Naess 1984, 1989; Porritt 1986; Bunyard and Morgan-Grenville 1987; Postel 1992) will find a very receptive audience in this region.

Considering the resurgence of Islam both as a political philosophy for governance and as a way of life, and its evident consistency with features of the growing environmental movement in the region, a deep analysis of the interaction between these two schools of thought may be necessary if we are to understand fully the future trends of water politics in the Middle East.

Notes

1 Despite enormous efforts made during the International Drinking Water Supply and Sanitation Decade of the 1980s, more than 1000 million people are still suffering from lack of access to adequate water supplies and at least 2 million children die every year because of unsafe water and inadequate sanitation. It is also predicted that by the year 2025, one third of the world's population will be suffering from chronic water scarcity (Bailey 1996: 1).
2 See for instance, Bulloch and Darwish 1993: 162; Butler 1995: 34; Haddad and Mizyed 1996: 3; Hillel 1994: 36; G. Joffe 1993: 65; Kliot 1994: V; Kolars and Mitchell 1991: 259; Mastrull 1995: 36.
3 Alam refers to this as the 'co-operative spillover', though she notes that it did not occur in the Indus Basin; 'Agreement on the Indus waters did not lead to talks and co-operation on wider Indo-Pakistani issues such as Kashmir, though that had been the initial hope of the intervenors. Co-operation, instead of snow balling, remained specific and muted' (1998: 263–4).
4 See for instance, Gleick 1993c: 108; Lipschutz 1989: chs 5–7; Lipschutz and Holdren 1990: 126; Shapland 1995: 305, 322), Waterbury 1994: 45.

Bibliography

Abate, Zewdie (1994) *Water Resources Development in Ethiopia*, Reading: Ithaca Press.

Abdel Mageed, Yahia (1994a) The central region: problems and perspectives. In: Rogers, P. and Lydon, P. (eds), *Water in the Arab World: Perspectives and Progress*, Cambridge, Mass.: Harvard University Press, pp. 101–119.

Abdel Mageed, Yahia (1994b) The Nile Basin: lessons from the past. In: Biswas, Asit K. (ed.), *International Waters of the Middle East: From Euphrates–Tigris to Nile*, Oxford University Press, pp. 156–84.

Abdulrazzak, Mohammed (1996) Strategies for water management in the GCC countries. Contribution to the Symposium on Water and Arab Gulf Development: Problems and Policies, University of Exeter.

Abramovitz, Janet N. (1996) *Imperilled Waters, Impoverished Future: The Decline of Freshwater Ecosystems*, paper no. 128, Washington DC: Worldwatch.

Achterberg, Wouter (1994) Can liberal democracy survive the environmental crisis? In: Zweers, Wim and Boersema, Jan J. (eds), *Ecology, Technology and Culture*, Cambridge: The White Horse Press, pp. 135–57.

Aczel, John (1991) Rising tide, *Water and Environment International*, 2 (13): 18–19, 40.

Adams, Robert Mc C. (1965) *Land Behind Baghdad: A History of Settlement on the Diyala Plains*. University of Chicago Press.

Adams, W. M. (1992) *Wasting the Rain: Rivers, People and Planning in Africa*, London: Earthscan.

Agnew, Clive (1995) Environmental change and environmental problems in the Middle East. In: Watkins, Eric (ed.), *The Middle Eastern Environment*, Cambridge: St Malo Press, pp. 21–34.

Agnew, Clive and Anderson, Ewan (1992) *Water Resources in the Arid Realm*, London and New York: Routledge.

Ahiram, Ephraim and Siniora, Hanna (1994) The Gaza Strip water problem – an emergency solution for the Palestinian population. In: Isaac, J. and Shuval, H. (eds) *Water and Peace in the Middle East*, Amsterdam and London: Elsevier, pp. 261–72.

Ahmad, Mahmood (1996) Sustainable water policies in the Arab region. Contribution to the Symposium on Water and Arab Gulf Development: Problems and Policies, University of Exeter.

Al-Alawi, Jamil and Abdulrazzak, Mohammed (1994) Water in the Arabian Peninsula: problems and perspectives. In: Rogers, P. and Lydon, P. (eds), *Water in the Arab World: Perspectives and Progress*, Cambridge, Mass.: Harvard University Press, pp. 171–202.

Alam, Undala Z. (1998) *Water Rationality: Mediating the Indus Treaty*, PhD Thesis, University of Durham.

Al-Kuwari, Ali K. (1996) Impact of the pattern of growth in oil economies on the depletion of water resources: the case of Qatar. Contribution to the Symposium on Water and Arab Gulf Development: Problems and Policies, University of Exeter.

Al-Ruwaih, F. M. (1995) Assessment and management of a desert basin in Kuwait, *Water International*, 20 (4): 213–24.
Al-Sheikh, Hamad M. H. (1996) Water resources and development in Saudi Arabia. Contribution to the Symposium on Water and Arab Gulf Development: Problems and Policies, University of Exeter.
Al-Tawati, Ali Hasan (1990) *Saudi Agricultural Policy from a Rural and Regional Perspective*. PhD Thesis, Syracuse University, US.
Al-Weshah, Radwan A. (1992) Jordan's water resources: technical perspective, *Water International*, 17 (3): 124–32.
Al-Zubari, Waleed K. (1996) Towards the establishment of a total water cycle management and re-use programme in the GCC countries. Contribution to the Symposium on Water and Arab Gulf Development: Problems and Policies, University of Exeter.
Allan, J. A. (1994a) Overall perspectives. In: Rogers, P. and Lydon, P. (eds), *Water in the Arab World: Perspectives and Progress*, Cambridge, Mass.: Harvard University Press, pp. 65–100.
Allan, J. A. (1994b) Economic and political adjustments to scarce water in the Middle East. In: Isaac, J. and Shuval, H. (eds), *Water and Peace in the Middle East*, Amsterdam and London: Elsevier, pp. 375–88.
Allan, J. A. (1994c) Water: a substitutable resource in the Middle East. Lecture at Durham University, 19 October.
Allan, J. A. (1995) Personal Communication, SOAS.
Allan, J. A. (1996) The political economy of water: reasons for optimism but long-term caution. In: Allan, J. A. with J. H. Court (eds), *Water, Peace and the Middle East*, London: I. B. Tauris, pp. 75–119.
Allan, J. A. with J. H. Court (eds) (1996) *Water, Peace and the Middle East: Negotiating Resources in the Jordan Basin*, London and New York: I. B. Tauris.
Allan, J. A. and Karshenas, M. (1996) Managing environmental capital: the case of water in Israel, Jordan, the West Bank and Gaza, 1947 to 1995. In: Allan, J. A. with J. H. Court (eds), *Water, Peace and the Middle East*, London: I. B. Tauris, pp. 121–33.
Allan, J. A. and Mallat, Chibli (1995) Introduction. In: Allan, J. A. et al. (eds), *Water in the Middle East: Legal, Political and Commercial Implications*, London and New York: I. B. Tauris, pp. 1–18.
Allan, J. A. and Mallat, Chibli with Shai Wade and Jonathan Wild (eds) (1995) *Water in the Middle East: Legal, Political and Commercial Implications*, London and New York: I. B. Tauris.
Allan, J. A. and Nicol, Alan (1996) The political economy of Middle East water: changing perceptions of the economic and strategic roles for water. Contribution to the Symposium on Water and Arab Gulf Development: Problems and Policies, University of Exeter.
Allan, J. A. Radwan L. (1996) *Perceptions of the Values of Water and Water Environments*. Proceedings of the European seminar on water geography. SOAS University of London and Middlesex University.
Allan, J. A. and Warren, A. (1993) *Deserts: the encroaching wilderness*, London: Mitchell Beazley.
Amir, Dror (1995) *The Environment and the Peace Process*. Israel Information Service Gopher, The Foreign Ministry, Information Division. URL: gopher://israel-inf.gov.il.

222 Bibliography

Anderson, Ewan W. (1991a) White Oil, *Geographical Magazine*, **63** (Feb): 10–14.
Anderson, Ewan W. (1991b) The source of power, *Geographical Magazine*, **63** (Mar): 12–15.
Anderson, Ewan W. (1991c) Making waves on the Nile, *Geographical Magazine*, **63**, (Apr): 10–13.
Anderson, Ewan W. (1991d) The violence of thirst, *Geographical Magazine*, **63**, (May): 31–34.
Anderson, Ewan W. (1995) Water Resources in the Middle East: Boundaries and Potential Legal Problems, Using Jordan as a Case Study. In: Blake, Gerald. H. *et al.* (eds), *The Peaceful Management of Transboundary Resources*, London: Graham Trotman, pp. 203–8.
Anderson, Kathleen and Holeman, Tim (1995) Lessons in conflict resolution: the case of the Denver Water Department. In: Dinar, Ariel and Loehman, Edna Tusak (eds), *Water Quantity/Quality Management and Conflict Resolution*, London: Praeger, pp. 233–47.
AOAD [Arab Organisation for Agricultural Development] (1996) Appropriate alternatives for irrigation management improvement in the Arab region. Contribution to the Symposium on Water and Arab Gulf Development: Problems and Policies, University of Exeter.
Arar, A. (1991) Wastewater reuse of irrigation in the Near East region, *Water Science and Technology WSTED4*, **23** (10/2): 2127–34.
Arlosoroff, S. (1996) Managing scarce water – recent Israeli experience. In: Allan, J. A. with J. H. Court (eds), *Water, Peace and the Middle East: Negotiating Resources in the Jordan Basin*, London and New York: I. B. Tauris, pp. 21–48.
Attfield, R. (1983) *The Ethics of Environmental Concern*, Oxford: Blackwell.
Bailey, Bob (1996) Sky TV Special Report. Sunday 25 August, 10:30 pm.
Bailey, Richard (ed.) (1996) *Water and Environmental Management in Developing Countries*, London: The Chartered Institution of Water and Environmental Management.
Bakour, Yahia and Kolars, John (1994) The Arab Mashrek: hydrologic history, problems and perspectives. In: Rogers, P. and Lydon, P. (eds), *Water in the Arab World: Perspectives and Progress*. Cambridge, Mass.: Harvard University Press, pp. 121–45.
Banks, P. A. (1991) Wastewater reuse case studies in the Middle East, *Water Science and Technology WSTED 4*, **23** (10/2): 2141–8.
Barham, John (1994) Demirel raises stakes in tense regional game: project means Turkish hand on Syrian and Iraqi water supply, *Financial Times*, 10 November: 8.
Barham, John (1995) Turkey's doubly shunned lick their wounds, *Financial Times*, 17 March: 2.
Barham, John (1996) Euphrates power plant generates new tension: A three-nation dispute over water, *Financial Times*, 15 February: 3.
Bassett, Libby (ed.) (1986) *Environment and Development: Opportunities in Africa and the Middle East* (Conference Summary), World Environment Center, New York.
Baumgartner, A. and Reichel, E. (1975) *The World Water Balance*, Munich: Elsevier.

Beaumont, Peter. (1978) The River Euphrates: an international problem of water development resources, *Environmental Conservation*, 5 (1): 35–44.
Beaumont, Peter, Blake, G. H. and Wagstaff, J. M. (1988), *The Middle East: A Geographical Study*, London: David Fulton.
Beaumont, Peter, Bonine, M., McLachlan, K. and McLachlan, A. (eds) (1989) *Qanat, Kariz and Khattara: Traditional water systems in the Middle East and North Africa*, The Centre for Middle Eastern Studies, School of Oriental & African Studies in association with Middle East & North African Studies Press, London.
Beaumont, Peter (1989) *Environmental Management and Development in Dry Lands*, London: Routledge.
Ben Meir, Meir (1994) Water management policy in Israel: a comprehensive approach. In: Isaac, J. and Shuval, H. (eds), *Water and Peace in the Middle East*, Amsterdam and London: Elsevier, pp. 33–40.
Ben-Shahar, Haim et al. (1989) *Economic Co-operation and Middle East Peace*, London: Weidenfeld & Nicolson.
Ben-Yamin, M. (1991) Israel's water crisis deepens, *World Water and Environmental Engineer*, (Mar): 29–30.
Bennett, O. (1991) *Greenwar: Environment and Conflict*, London: Panos.
Berber, F. F. (1959) *Rivers in International Law*, London: Stevens & Sons.
Berkes, Fikret (ed.) (1989) *Common Property Resources: Ecology and Community-based Sustainable Development*, London: Belhaven.
Berkoff, Jeremy (1994) *A Strategy for Managing Water in the Middle East and North Africa*, Washington, DC: The World Bank.
Beschorner, Natasha (1992) *Water and Instability in the Middle East*, Adelphi Paper 273, London: International Institute for Strategic Studies.
Bhatia, Ramesh, Cestti, Rita and Winpenny, James (1993) *Policies for Water Conservation and Reallocation: Good Practice Cases in Improving Efficiency and Equity*, Washington, DC: World Bank.
Bilen, Ozden (1994) Prospects for technical cooperation in the Euphrates–Tigris basin. In: Biswas, Asit K. (ed.), *International Waters of the Middle East: From Euphrates–Tigris to Nile*, Oxford University Press, pp. 95–116.
Bingcai, Wei (1994) Excerpts from seminar discussions. In: Sun, Peter (ed.), *Multipurpose River Basin Development in China*, Washington, DC: World Bank, pp. 23–9.
Biswas, Asit K. (ed.) (1978) *United Nations Water Conference: Summary and Main Documents*, Oxford: Pergamon.
Biswas, Asit K. (1983) Some major issues in river basin management for developing countries. In: Zaman, M. et al. (eds), *River Basin Development*, Dublin: Tycooly, pp. 17–27.
Biswas, Asit K. (1991a) Global warming, *World Water*, (Jul/Aug): 36–7.
Biswas, Asit K. (1991b) Water resources in the 21st century, *Water International*, 16 (3): 142–4.
Biswas, Asit K. (1992) Indus water treaty: the negotiating process, *Water International*, 17 (4): 201–9.
Biswas, Asit K. (ed.) (1994) *International Waters of the Middle East: From Euphrates–Tigris to Nile*, Oxford University Press.
Blackaby, Frank T. and Tolba, Mostafa (1986) Preface. In Westing, Arthur H. (ed.), *Global Resources and International Conflict: Environments Factors in Strategic Policy and Action*, Oxford University Press.

224 Bibliography

Blake, Gerald H., Hildesley, William J., Pratt, Martin A., Ridley, Rebecca J. and Schofield, Clive H. (eds) (1995) *The Peaceful Management of Transboundary Resources*, London: Graham & Trotman.

Blom, G. (1992) The Netherlands policy on transboundary river basin management, *European Water Pollution Control*, 2 (3): 20–2.

Blomquist, William (1992) *Dividing the Waters: Governing Groundwater in Southern California*, Institute for Contemporary Studies Press, San Francisco, CA.

Bodgener, Jim (1984) Oman develops skills ancient and modern, *Middle East Economic Digest*, 10 August: 38.

Bodgener, Jim (1994) Pouring water on to arid plains, *Middle East Economic Digest*, 25 November: 22.

Boersema, Jan J. (1994) First the Jew but also the Greek: in search of the roots of the environmental problem in Western civilisation. In: Zweers, Wim and Boersema, Jan J. (eds), *Ecology, Technology and Culture*, Cambridge: White Horse Press, pp. 20–55.

Bookchin, M. (1986) *The Modern Crisis*, Philadelphia: New Society.

Brennan, A. (1988) *Thinking About Nature*, London: Routledge.

Brown, Lester R. (1977) Redefining national security, Worldwatch paper no. 14, Washington DC: Worldwatch Institute.

Brown, Lester R. (1986) Redefining national security. In: Brown, L. R. *et al.* (eds), *State of the World 1986*, New York: Norton, pp. 195–211.

Brown, Neville (1989) Climate, ecology and international security, *Survival*, 31 (6): 519–32.

Brown, Neville (1990) Planetary geopolitics, *Millennium: Journal of International Studies*, 19 (3): 447–60.

Bruhacs, J. (1992) Evaluation of the legal aspects of projects in international rivers, *European Water Pollution Control*, 2 (3): 10–13.

Bulloch, John (1993) Troubled Waters: Middle East battles will be fought not over oil but the vital resources of three great river systems, *Independent on Sunday*, 14 November: 12.

Bulloch, John and Darwish, Adel (1993) *Water Wars: Coming Conflicts in the Middle East*, London: Victor Gollancz.

Bunyard, P. and Morgan-Grenville, F. (eds) (1987) *The Green Alternative*, London: Methuen.

Burchi, Stefano (1992) Legal aspects of planning for transboundary river basin management and conservation, *European Water Pollution Control*, 2 (3): 14–19.

Burdon, D. J. (1982) Hydrogeological conditions in the Middle East, *Quarterly Journal of Engineering Geology*, 15 (2): 71–82.

Butler, Daniel (1995) A world in hot water, *Accountancy*, 116 (1228): 34–8.

Buzan, Barry (1991) *People, States and Fear*, 2nd edn, Hemel Hempstead: Harvester Wheatsheaf.

Caldwell, Lynton Keith (1990) *International Environmental Policy: Emergence and Dimensions*, 2nd edn, Durham: Duke University Press.

Calvert, Peter (1993) Water politics in Latin America. In: Thomas, Caroline and Howlett, Darryl (eds), *Resource Politics: Freshwater and Regional Relations*, Buckingham: Open University Press, pp. 47–64.

Cano, Guillermo J. (1986) The 'Del Plata' Basin: summary chronicle of its development process and related conflicts. In: Vlachos, E. *et al.* (eds), *The Management of International River Basin Conflicts*, Washington, DC: George Washington University, ch. 2.3: 70pp.

Cano, Guillermo J. (1989) The development of the law of international water resources and the work of the International Law Commission, *Water International*, **14** (4): 167–71.

Caponera, Dante A. (1973) *Water Law in Moslem Countries*, Rome: FAO.

Caponera, Dante A. (1983) International river law. In: Zaman, M. *et al.* (eds), *River Basin Development*, Dublin: Tycooly, pp. 173–84.

Caponera, Dante A. (1992) *Principles of Water Law and Administration, National and International*, Rotterdam: A. A. Balkami.

Caponera, Dante A. (1995) Shared waters and international law. In: Blake, Gerald. H. *et al.* (eds), *The Peaceful Management of Transboundary Resources*, London: Graham & Trotman, pp. 121–6.

Capra, F. (1985) *The Turning Point*, London: Flamingo.

Chalabi, Hasan and Majzoub, Tarek (1995) Turkey, the waters of the Euphrates and public international law. In: Allan, J. A. *et al.* (eds), *Water in the Middle East: Legal, Political and Commercial Implications*, London and New York: I. B. Tauris, pp. 189–238.

Charnock, Anne (1984) Water Resources, *Middle East Economic Digest*, 10 August: 27–37.

Chesnoff, R. Z. (1988) Water feeds the flames, *US News and World Report*, 21 November.

Chomchai, Prachoom (1986) The Mekong Committee: an exercise in regional co-operation to develop the lower Mekong Basin. In: Vlachos, E. *et al.* (eds) *The Management of International River Basin Conflicts*, Washington, DC: George Washington University, ch. 2.4: 36pp.

Chomchai, Prachoom (1995) Management of transboundary water resources: a case study of the Mekong. In: Blake, Gerald. H. *et al.* (eds), *The Peaceful Management of Transboundary Resources*, London: Graham & Trotman, pp. 245–60.

Choudhury, G. R. and Khan, T. A. (1983) Developing the Ganges basin. In: Zaman, M. *et al.* (eds), *River Basin Development*, Dublin: Tycooly, pp. 28–39.

Clarke, Robin (1991) *Water: The International Crisis*, London: Earthscan.

Cooley, John K. (1984) War Over Water, *Foreign Policy*, (54): 3–26.

Cotgrove, Stephen (1982) *Catastrophe & Cornucopia: The Environment: Politics and the Future*, Chichester, Jussex, John Wiley.

Dabbagh, T., Sadler, P., Al-Saqabi, A. and Sadeqi, M. (1994) Desalination: an emergent option. In: Roger, P. and Lydon, P. (eds), *Water in the Arab World: Perspectives and Prognoses*, Cambridge, Mass.: Harvard University Press, pp. 203–42.

Dabelko, Geoffrey D. and Dabelko, David D. (1995) *Environmental Security: Issues of Conflict and Redefinition*, Woodrow Wilson Center's Environmental Change and Security Project Report, Issue 1 (Spring).

Dalby, Simon (1992a) Ecopolitical discourse: environmental security and political geography, *Progress in Human Geography*, **16**: 503–522.

Dalby, Simon (1992b) Security, modernity, ecology: the dilemmas of post-Cold War security discourse, *Alternatives*, **7** (1): 95–134.

Dalby, Simon (1995) The threat from the South? Global justice and environmental security. In: Deudney, Daniel and Matthew, Richard (eds), *Contested Ground: Security and Conflict in the New Environmental Politics*, Albany NY: SUNY Press, pt III, ch. 3.

Dana, Andrew (1990) Water resource management. In: Baden, John. A and Donald, Leaf (eds), *The Yellowstone Primer*, San Francisco: Pacific Research Institute for Public Policy, pp. 49–79.

Dave K. G. (1991) Water conservation as a tool to fight water scarcity – a case study, *Journal of Indian Waterworks Association*, **23** (1): 5–9.

Davey, Tom (1985) Water – a global crisis in the making?, *Water and Pollution Control*, **123** (6): 10–18.

de Jong, Remy L. (1989) Water resources of the GCC: international aspects, *Journal of Water Resources Planning and Management*, **115** (4): 503–10.

de-Shalit, Avner (1995) From the political to the objective: the dialectics of Zionism and the environment, *Environmental Politics*, **4** (1): 70–87.

Dellapenna, Joseph (1995) Building international water management institutions: the role of treaties and other legal arrangements. In: Allan, J. A. et al. (eds), *Water in the Middle East: Legal, Political and Commercial Implications*, London and New York: I. B. Tauris, pp. 55–89.

deShazo, Randal and Sutherlin, John W. (1994) *Reassessing the Middle Eastern 'Peace Pipeline' in the aftermath of the Gaza–Jericho Agreement*, Environmental Institute, University of New Orleans.

Deudney, Daniel (1990) The case against linking environmental degradation and national security, *Millennium: Journal of International Studies*, **19** (3): 461–76.

Deudney, Daniel (1991) Environment and security: muddled thinking, *Bulletin of the Atomic Scientists*, **47** (3): 22–8.

Deudney, Daniel and Matthew, Richard (eds) (1995) *Contested Ground: Security and Conflict in the New Environmental Politics*, Albany, NY: SUNY Press.

Dillman, J. (1989) Water rights in the Occupied Territories, *Journal of Palestine Studies*, **19** (1): 46–71.

Dinar, Ariel and Loehman, Edna Tusak (eds) (1995) *Water Quantity/Quality Management and Conflict Resolution*, London: Praeger.

Dobson, Andrew (1990) *Green Political Thought: An Introduction*, London: Unwin Hyman.

Dolatyar, Mostafa (1995) Water diplomacy in the Middle East. In: Watkins, Eric (ed.) *The Middle Eastern Environment*, Cambridge: St Malo Press, pp. 35–43.

Dolatyar, Mostafa (1996) The roots of environmental crises: an Islamic perspective, *Iranian Journal of International Affairs*, **8** (4): 761–71.

Downey, Terrence J. and Mitchel, Bruce (1993) Middle East water: acute or chronic problem?, *Water International*, **18** (1): 1–4.

Drake, C. (1988) Oman: tradition and modern adaptations to environment, *Focus*, (Summer issue): 15–20.

du Bois, François (1995) Water law in the economy of nature. In: Allan, J. A. et al. (eds), *Water in the Middle East: Legal, Political and Commercial Implications*, London and New York: I. B. Tauris, pp. 111–23.

Duma, C. (1988) Turkey's peace pipeline. In: Starr, Joyce R. and. Stoll, Daniel C. (eds), *The Politics of Scarcity: Water in the Middle East*, Washington DC: Center for Strategic and International Studies, pp. 119–24.

Duwais, Aboul-Aziz Mohammed (1990) *The Saudi Agricultural Sector Model: Structure and Policy Applications*, PhD Thesis, Oklahoma State University, US.
Dyer, Hugh C. (1996) Environmental security as a universal value: implications for international theory. In: Vogler, John and Imber, Mark F. (eds), *The Environment and International Relations*, London and New York: Routledge, pp. 22–40.
Dyer, J. A. (1989) A Canadian approach to drought monitoring for famine relief in Africa, *Water International*, 14 (4): 198–205.
Eaton, J. W., and Eaton, D. J. (1994) Water utilization in the Yarmuk-Jordan, 1192–1992. In: Isaac, J. and Shuval, H. (eds), *Water and Peace in the Middle East*, Amsterdam and London: Elsevier, pp. 93–106.
Eckersley, Robyn (1992) *Environmentalism and Political Theory: Toward an Ecocentric Approach*, London: UCL Press.
Economist, The (1987) Where dams can cause wars, *The Economist*, 18 July: 57–8.
Economist, The (1990) Water wars in the Middle East, *The Economist*, 12 May: 54–9.
ECSCWA (1994) *Land and Water Policies in the Arab Region*, Contribution to the expert group consultation on sustainable agricultural and rural development (SARD), Cairo, 25–29 September 1994. ECSCWA and FAO, E/ESCWA/AGR/1994/2.
ECWA (1978) Report on the Economic Commission for Western Asia, Regional Preparatory Meeting for the UNWC. In: *UNWC, Water Management and Development, Proceedings of the United Nations Water Conference, Mar del Plata, Argentina March 1977*, New York: Pergamon, vol. 1, pt 2, pp. 639–58.
Ekins, P. (ed.) (1986) *The Living Economy*, London: Routledge & Kegan Paul.
El Morr, Awad (1995) Water resources in the Middle East: some guiding principles. In: Allan, J. A. et al. (eds.) *Water in the Middle East: Legal, Political and Commercial Implications*, London and New York: I. B. Tauris, pp. 293–300.
El-Nashar, A. M. and Qamhiyeh, A. A. (1991) Performance and reliability of a solar desalination plant during a five-year operating period, *Desalination*, (82): 165–74.
Elhance, Arun (1995) Geography and hydropolitics. In: Deudney, Daniel and Matthew, Richard (eds), *Contested Ground: Security and Conflict in the New Environmental Politics*, Albany, NY: SUNY Press, part III, ch. 2.
Elliot, R. and Gare, A. (eds) (1983) *Environmental Philosophy*, Milton Keynes: Open University Press.
Elliott, Michael (1991) Water wars, *Geographical Magazine*, 63, May: 28–30.
Elmer-Dewitt, Philip (1989) Preparing for the worst, endangered Earth, *TIME*, 2 January.
Elmusa, Sharif S. (1994) Towards an equitable distribution of the common Palestinian-Israeli waters: an international water law framework. In: Isaac, J. and Shuval, H. (eds) *Water and Peace in the Middle East*, Amsterdam and London: Elsevier, pp. 451–68.
Encarta (1994) *Encarta Multimedia Encyclopedia*, Microsoft Corporation, California.
Engelmann, Kurt E. (1994) Amu Darya, A contribution to *Encarta Multimedia Encyclopedia*, Microsoft Corporation, California.

Fahim, Hussein M. (1996) Social sciences and the crafting of water policy. Contribution to the Symposium on Water and Arab Gulf Development: Problems and Policies, University of Exeter.

Falkenmark, Malin (1984) Water: the silent messenger between cause and effect in environmental problems, *Water International*, **9** (2): 62–5.

Falkenmark, Malin (1986a) Fresh water as a factor in strategic policy and action. In: Westing, Arthur H. (ed.), *Global Resources and International Conflict*, Oxford University Press, pp. 85–113.

Falkenmark, Malin (1986b) Fresh water: time for a modified approach, *Ambio*, **15** (4): 192–200.

Falkenmark, Malin (1989) Middle East hydropolitics: water scarcity and conflicts in the Middle East, *Ambio*, **18** (6): 350–2.

Falkenmark, Malin (1990) Global water issues confronting humanity, *Journal of Peace Research*, **27** (2): 177–90.

Falkenmark, Malin and Widstrand, Carl (1992) Population and water resources: a delicate balance, *Population Bulletin*, **47** (3), Washington, DC: Population Reference Bureau, Inc.

FAO (1978) *Systematic Index of International Water Resources Treaties, Declarations, Acts and Cases by Basin*, Legislative study no. 15, Rome: FAO.

FAO (1979) *Survey and Evaluation of Available Data on Shared Water Resources in the Gulf States and the Arabian Peninsula*, Rome: FAO.

Feitelson, Eran (1996) The implications of changes in perceptions of water in Israel for peace negotiations with Jordan and Palestinians. In: Allan, J. A. and Radwan L. (eds), *Perceptions of the Values of Water and Water Environments*, Proceedings of the European seminar on water geography. London SOAS and Middlesex University, pp. 17–21.

Feitelson, Eran. and Haddad, M. (eds) (1994) *Joint Management of Shared Aquifers*, Jerusalem: Harry S. Truman Research Institute, Hebrew University and the Palestinian Consultancy Group.

FitzGibbon, J. E. (ed.) (1990) *International and Transboundary Water Resources Issues*, Symposium Papers, American Water Resources Association (AWRA), Bethesda, Canada.

Fleckseder, H. (1992) Rhine–Danube-Project: lessons to be learnt in the fields of water quality, *European Water Pollution Control*, **2** (3): 32–6.

Flint, Courtney G. (1995) Recent developments of the International Law Commission regarding International Watercourses and their implications for the Nile river, *Water International*, **20** (4): 197–204.

Fox, I. K. and LeMarquand, D. (1979) International river basin co-operation: the lessons from experience, *Water Supply and Management*, **3** (1): 9–27.

Fox, Warwick (1984) Deep ecology: a new philosophy of our time? *The Ecologist*, **14** (5/6): 194–200.

Frankel, N. (1991) Water and Turkish Foreign Policy, *Political Communication and Persuasion*, (8): 257–311.

Frankel, N. (1992) Water and politics: the Turkish perspective, *Middle East Focus* (Spring): 4–8, 17.

Frey, Frederick W. (1993a) Power, conflict and co-operation, *Research and Exploration*, **9** (special issue): 19–37.

Frey, Frederick W. (1993b) The political context of conflict and co-operation over international river basins, *Water International*, **18** (1): 54–68.

Frey, Frederick, W. and Naff, Thomas (1985) Water: an emerging issue in the Middle East? *Annals of the American Academy of Political Scientists*, (482): 65–84.

GAP Main Web (1996a) About GAP: General Information. URL: http://urfa.gap.gov.tr/.

Garretson, Albert Henry, Hayton, R. D. and Olmstead, C. J. (eds) (1967) *The Law of International Drainage Basins*, published for the Institute of International Law, New York University School of Law, Dobbs Ferry, NY: Oceana.

Gelsse, Monica G. and Arenas, Hernan S (1995) Chile-Bolivia Relations: The Lauca River Water Resources. In: Blake, Gerald. H. *et al.* (eds) *The Peaceful Management of Transboundary Resources*, London: Graham & Trotman, pp. 227–86.

Gibbons, Diana C. (1986) *The Economic Value of Water*, Washington, DC: Resources for the Future.

Gischler, Christiaan E. (1979) *Water in the Arab Middle East and North Africa*, Cambridge: Middle East and North African Studies Press.

Gleick, Peter H. (1989a) Climate change and international politics: problems facing developing countries, *Ambio*, **18** (6): 333–9.

Gleick, Peter H. (1989b) The implications of global climatic changes for international security, *Climatic Change*, (15): 309–25.

Gleick, Peter H. (1991) Environment and security: the clear connections, *Bulletin of the Atomic Scientists*, **47** (3): 17–21.

Gleick, Peter H. (1993a) Water and conflict: fresh water resources and international security, *International Security*, **18** (1): 79–112.

Gleick, Peter H. (ed.) (1993b) *Water in Crisis: A Guide to the World's Fresh Water Resources*, Oxford University Press.

Gleick, Peter H. (1993c) Water in the 21st century. In: Gleick, Peter H., (ed.), *Water in Crisis: A Guide to the World's Fresh Water Resources*, Oxford University Press, pp. 105–13.

Gleick, Peter H. (1994a) Water, war and peace in the Middle East, *Environment*, **36** (3): 6–15, 35–42.

Gleick, Peter H. (1994b) Reducing the risks of conflict over fresh water resources in the Middle East. In: Isaac, J. and Shuval, H. (eds), *Water and Peace in the Middle East*, Amsterdam and London: Elsevier, pp. 41–54.

Goldberg, David (1995) World Bank Policy on Projects on International Waterways in the Context of Emerging International Law and the Work of the International Law Commission. In: Blake, Gerald. H. *et al.* (eds) *The Peaceful Management of Transboundary Resources*, London: Graham & Trotman, pp. 153–66.

Golubev, G. N. (1990) Economic activity, water resources and the environment, *Hydrological Science Journal*, **28** (1): 57–75.

Gosh, Hilary (1997) Changing directions in Saudi Arabia, *Reuters World Report*, 28 July, Dubai.

Gowers, Andrew and Walker, Tony (1989) Middle East Fears War of Parched Throats, *Financial Times*, 21 February: 7.

Grolier (1993) *The New Grolier Multimedia Encyclopedia*, Release 6, Grolier Inc., USA.

Grove, A. T. (1985) *The Niger and Its Neighbours: Environmental History and Hydrobiology of the Major West African Rivers*, Rotterdam and Boston: A. A. Balkema.

Grove-White, Robin (1993) Environmentalism: a new moral discourse for technological society? In: Milton, Kay (ed.), *Environmentalism: The View from Anthropology*, London and New York: Routledge, pp. 18–30.

Gruen, George E. (1991) *The Water Crisis: The Next Middle East Crisis?*, Los Angeles: Simon Wiesenthal Center.

Guest, Anne (1993) Sub-Saharan Africa. In: Thomas, Caroline and Howlett, Darryl (eds), *Resource Politics*, Buckingham: Open University Press, pp. 129–48.

Gunter, M. M. (1990) The Kurds in Turkey. In: *The New Grolier Multimedia Encyclopedia*, Release 6, Grolier Inc, USA.

Gup, Ted (1993) Clinton says it's not just the economy that matters, *TIME*, 12 July: 38.

Gurr, Ted Robert (1985) On the political consequences of scarcity and economic decline, *Internationial Studies Quarterly*, 29 (1): 51–75.

Guttman, Cynthia (1966) The call of the desert, *UNESCO Sources*, p. 9. http://www2.elibrary.com.

Haddad, M. and Mizyed, N. (1996) Water Resources in the Middle East: conflict and solutions. In: Allan, J. A. and Court, J. H. (eds), *Water, Peace and the Middle East: Negotiating Resources in the Jordan Basin*, London: I. B. Tauris, pp. 3–17.

Haddadin, M. J. (1996) Water management: a Jordanian viewpoint. In: Allan, J. A. with J. H. Court (eds), *Water, Peace and the Middle East: Negotiating Resources in the Jordan Basin*, London and New York: I. B. Tauris, pp. 59–73.

Hamdy, Atef et al. (1995) Water crisis in the Mediterranean: agricultural water demand management, *Water International*, 20 (4): 176–87.

Hardan, Adai (1993) Sharing the Euphrates: Iraq, *Research and Exploration*, 9 (special issue): 73–9.

Hartvelt, Frank and Okun, Daniel A. (1991) Capacity building for water resources management, *Water International*, 16 (3): 176–83.

Hassan, S. (1991) Environmental issues and security in South Asia, *Adelphi Paper* no. 262: 3–69.

Hatami, Haleh and Gleick, P. H. (1993) *Chronology of Conflict over Water in the Legends, Myths, and History of the Ancient Middle East*, Oakland, California: Pacific Institute for Studies in Development, Environment, and Security.

Hatami, Haleh and Gleick, P. H. (1994) Conflicts over water in the myths, legends, and ancient history of the Middle East, *Environment*, 36 (3): 10–11.

Hayton, Robert D. (1983) Law of international water resources systems. In: Zaman, M. et al. (eds) *River Basin Development*, Dublin: Tycooly, pp. 195–211.

Hayton, Robert D. and Utton, Albert E. (1989) Transboundary groundwaters: the Bellagio Draft Treaty, *Natural Resource Journal*, 29 (3): 663–721.

Hewedy, Amin (1989) *Militarization and Security in the Middle East: Its Impact on Development and Democracy*, Tokyo: UN University.

Hillel, Daniel (1994) *Rivers of Eden – The Struggle for Water and the Quest for Peace in the Middle East*, Oxford University Press.

Hirshleifer, Jack et al. (1969) *Water Supply: Economics, Technology, and Policy*, University of Chicago Press.

Hitti, Philip K. (1970) *History of the Arabs: From the Earliest Times to the Present*, New York: St. Martins Press.

Homer-Dixon, Thomas F. (1990) Environmental change and violent conflict, American Academy of Arts and Sciences, occasional paper no. 4.

Homer-Dixon, Thomas F. (1991a) On the threshold: environmental changes as causes of acute conflict, *International Security*, **16** (2): 76–116.
Homer-Dixon, Thomas F. (1991b) Environmental change and acute conflict: a research agenda. Global Environmental Change Committee, Social Science Research Council, New York. pp. 1–47.
Homer-Dixon, Thomas F. (1992) Population growth and conflict. In: *Environmental Dimensions of Security*, Proceedings from a AAAS Annual Meeting Symposium.
Homer-Dixon, Thomas F. (1994a) Population and conflict. Prepared for the IUSSP Distinguished Lecture Series, International Conference on Population and Development Cairo, 1994.
Homer-Dixon, Thomas F. (1994b) Environmental scarcities and violent conflict: evidence from cases, *International Security*, **19** (1): 5–40.
Homer-Dixon, Thomas F. (1995) *Strategies for Studying Causation in Complex Ecological-Political Systems*, The Peace and Conflict Studies Program at University College, Toronto.
Hotten, Russell (1995) Yorkshire water dispute escalates, *Independent*, 6 May: 17.
Howell, P. P. and Allan, J. A. (eds) (1994) *The Nile: Sharing a Scarce Resource*, Cambridge University Press.
Hurewitz, J. (1956) *Diplomacy in the Near and Middle East*, vol. 2. Princeton: D. V. Nostrand.
Hvidt, Martin (1997) *Water, Technology and Development: Upgrading Egypt's Irrigation System*, London I. B.: Tauris.
Hyde-Price, Adrian (1993) Eurasia. In: Thomas, Caroline and Howlett, Darryl (eds), *Resource Politics*, Buckingham: Open University Press, pp. 149–70.
Ibn Manzur (1959) *Lisan al-'Arab*, Vol. 3, Beirut.
ILA (1967) *Helsinki Rules on the Uses of the Waters of International Rivers*, London: International Law Association.
ILC (1991) Report of the International Law Commission on the work of its forty-third session, New York: United Nations.
Imber, Mark F. (1991) Environmental security: a task for the UN system, *Review of International Studies*, (17): 201–12.
IRAL (1984) International Law commission: new report on international rivers law, *International Rivers and Lakes*, (4): 3–4.
IRAL (1985) Report on law of non-navigational watercourses, *International Rivers and Lakes*, (5): 13–15.
Irani, Rustam (1991) Water wars, *New Statesman and Society*, **4**, 3 May: 24–5.
Irvine, S. and Ponton, A. (1988) *A Green Manifesto: Policies for a Green Future*, London: MacDonald Optima.
Isaac, Jad and Shuval, Hillel (eds) (1994) *Water and Peace in the Middle East*, First Israeli–Palestinian International Academic Conference on Water (December 1992: Zurich, Switzerland), Amsterdam and London: Elsevier.
Jacobs, Frans (1994) Can liberal democracy help us to survive the environmental crisis? In: Zweers, Wim and Boersema, Jan J. (eds), *Ecology, Technology and Culture*, Cambridge: White Horse Press, pp. 158–63.
Jayal N. D. (1985) Destruction of water resources – the most critical ecological crisis of East Asia, *Ambio*, **14** (2): 95–8.
Jellali, Mohammed and Jebali, Ali (1994) Water resources development in the Maghreb countries. In: Rogers, P. and Lydon, P. (eds), *Water in the*

Arab World: Perspectives and Progress, Cambridge, Mass.: Harvard University Press, pp. 147–70.

Joffe, E. G. H. (1985) Agricultural development in Saudi Arabia: the problematic path to self-sufficiency. In: Beaumont, P. and McLachlan, K. (eds), *Agricultural Development in the Middle East,* New York: John Wiley, pp. 209–25.

Joffe, George (1993) The issue of water in the Middle East and North Africa. In: Thomas, Caroline and Howlett, Darryl (eds), *Resource Politics: Freshwater and Regional Relations,* Buckingham: Open University Press, pp. 65–85.

Johnson, T. P. (1991) Writing for International Security: a contributors' guide, *National Security,* 16 (2): 172.

Joma, Hasam Addin Abdul Salam (1991) *The Earth as a Mosque: Integration of the Traditional Islamic Environmental Planning Ethic with Agricultural and Water Development Policies in Saudi Arabia,* PhD Thesis, University of Pennsylvania, USA.

Kally, Elisha (1986) *A Middle East Water Plan under Peace,* Armand Hammer Fund for Economic Cooperation, Tel Aviv University, Tel Aviv.

Kally, Elisha and Fishelson, Gideon (eds) (1993) *Water Resources and the Arab-Israeli Peace Process,* London: Praeger.

Katko, Tapio S. (1990) Cost recovery in water supply in developing countries, *Water Resources Development,* 6 (2): 86–94.

Keen, M. (ed.) (1988) *Arab Agriculture 1987,* Bahrain: Falcon.

Kemp, Geoffrey (1978). Scarcity and strategy, *Foreign Affairs,* 57 (Jan): 396–414.

Kemp, Peter (1995) Adopting new approaches as deficits loom, *Middle East Economic Digest,* Special Report. (27 Jan): 8–9.

Khatib, Hisham (1996) Planning viable water strategies for the Arabian Peninsula. Contribution to the Symposium on Water and Arab Gulf Development: Problems and Policies, University of Exeter.

Khordagui, Hosny (1996) Prospects of non-conventional water resources in the Arabian Peninsula. Contribution to the Symposium on Water and Arab Gulf Development: Problems and Policies, University of Exeter.

Kirk, Elizabeth J. (1991) The greening of security: environmental dimensions of national, international, and global security after the Cold War. American Association for the Advancement of Science, Washington DC.

Kirmani, Syed S. (1990) Water, peace and conflict management: the experience of the Indus and Mekong river basins, *Water International,* 15 (4): 200–5.

Kirmani, Syed S. and Rangeley, Robert (1994) *International Inland Waters: Concepts for a More Active World Bank Role,* Washington, DC: World Bank.

Kiss, Alexandre (1992) Legal aspects of polluted sediments, *European Water Pollution Control,* 2 (3): 7–9.

Kliot, Nurit (1994) *Water Resources and Conflict in the Middle East,* London and New York: Routledge.

Kliot, Nurit (1995) Building a legal regime for the Jordan–Yarmuk river system: lessons from other international rivers. In: Blake, Gerald. H. *et al.* (eds), *The Peaceful Management of Transboundary Resources,* London: Graham & Trotman, pp. 187–202.

Kneese, Alan V. (1984) *Measuring the Benefits of Clean Air and Water,* Washington, DC: Resources for the Future.

Kolars, John (1991) The Future of the Euphrates River, World Bank Conference on Comprehensive Water Resources Management Policy. Washington, DC, June, pp. 1–29.

Kolars, John (1992) Water resources of the Middle East, *Canadian Journal of Development Studies*, Special Issue: Sustainable Water Resources Management in Arid Countries: 103–19.

Kolars, John (1993) The Middle East's growing water crisis, *Research and Exploration*, 9 (Special issue): 39–49.

Kolars, John (1994) Problems of international river management: the case of the Euphrates. In: Biswas, Asit K. (ed.), *International Waters of the Middle East: From Euphrates–Tigris to Nile*, Oxford University Press, pp. 44–94.

Kolars, John F. and Mitchell, William A. (1991) *The Euphrates River and the Southeast Anatolia Development Project*, Carbondale, IL: Southern Illinois University Press.

Krishna, Raj (1995) International watercourses: World Bank experience and policy. In: Allan, J. A. et al. (eds), *Water in the Middle East: Legal, Political and Commercial Implications*, London and New York: I. B. Tauris, pp. 29–54.

Lacayo, Richard (1991) A war against the Earth, *TIME*, 4 February 32–4.

Le Moigne, Guy (1994a) Policy issues and World Bank experience in multipurpose river basin development. In: Sun, Peter (ed.), *Multipurpose River Basin Development in China*, Washington, DC: World Bank, pp. 7–16.

Le Moigne, Guy (1994b) *A Guide to the Formulation of Water Resources Strategy*, World Bank Technical paper no. 263, Washington, DC: World Bank.

Le Moigne, Guy et al. (1994) *Water Policy and Water Markets*, selected papers and proceedings from the World Bank's Ninth Annual Irrigation and Drainage Seminar, Annapolis, Maryland, December 8–10, 1992; World Bank Technical paper no. 249.

Lean Geoffrey (1993) Troubled Waters, *Observer*, 4 July: 16–23, 25.

Lean Geoffrey (1995a) Global warming is leading to climatic upheaval, *Independent on Sunday*, 15 October.

Lean, Geoffrey (1995b) Global Warming puts Britain's climate down the plughole, *Independent on Sunday*, 26 March: 18.

Lean, Geoffrey (1995c) Sea levels rising twice as fast as feared, *Independent on Sunday*, 26 February.

Lean, Geoffrey (1995d) The food runs out: the world's cupboard is bare, *Independent on Sunday*, 12 November 17.

Lee, K. (1989) *Social Philosophy and Ecological Scarcity*, London: Routledge.

LeMarquand, David G. (1986) International development of the Senegal River. In: Vlachos, E., et al. (eds), *The Management of International River Basin Conflicts*, Washington, DC: George Washington University, ch. 2.2: 47pp.

LeMarquand, David G. (1990) International development of the Senegal River, *Water International*, 15 (4): 223–30.

Lemonick, Michael D. (1991) Dead sea in the making, *TIME*, 11 February: 40–2.

Leopold, A. (1949) *A Sand County Almanac*, Oxford University Press.

Libiszewski, Stephanie (1995) *Water Disputes in the Jordan Basin Region and their Role in the Resolution of the Arab–Israeli Conflict*, ENCOP occasional paper no. 13, Center for Security Studies and Conflict Research and Swiss

Peace Foundation, Zurich and Berne. Internet Version, URL: http://www.fsk.ethz.ch/encop/13/en13.htm.

Lichtenthaler, Gerhard (1996) Tribes and trends: changing perceptions of the value of water in Yemen. In: Allan, J. A. and Radwan L. (eds), *Perceptions of the Values of Water and Water Environments*, Proceedings of the European seminar on water geography, London SOAS and Middlesex University, pp. 121–25.

Linden, Eugene (1990) The last drops, *TIME*, 20 August: 58–61.

Linden, Eugene (1991) Making room for a stream of new arrivals has pushed nature to the wall, *TIME*, 18 November: 83–7.

Lipschutz, Ronnie D. (1989) *When Nations Clash: Raw Materials, Ideology and Foreign Policy*, New York and Cambridge: Ballinger.

Lipschutz, Ronnie D. (1992) What resources will matter? Environmental degradation as a security issue, Washington, DC: American Association for the Advancement of Science.

Lipschutz, Ronnie D. and Holdren, John P. (1990) Crossing borders: resource flows, the global environment, and international security, *Bulletin of Peace Proposals*, 21 (2): 121–33.

London, James B. and Miley, Harry W. Jr (1990) The interbasin transfer of water: an issue of efficiency and equity, *Water International*, 15 (4): 231–5.

Lonergan, Steve (1997) Environment and society in the Middle East: conflicts over water. In: Redclift, Michael and Woodgate, Graham (eds), *The International Handbook of Environmental Sociology*, Cheltenham: Edward Elgar, pp. 418–31.

Lowi, Miriam R. (1984) *The Politics of Water: the Jordan River and the Riparian States*, Quebec: McGill University.

Lowi, Miriam R. (1993a) *Water and Power: The Politics of a Scarce Resource in the Jordan River Basin*, Cambridge University Press.

Lowi, Miriam R. (1993b) Bridging the divide: transboundary resource disputes and the case of West Bank Water, *International Security*, 18 (1): 113–38.

Lowi, Miriam R. (1995) Rivers of conflict, rivers of peace, *Journal of International Affairs*, 49 (1): 123–44.

Lundqvist, Jan (1996) The triple squeeze on water: rain water, provided water and wastewater in socio-economic and environmental systems. In: Allan, J. A. and Radwan L. (eds), *Perceptions of the Values of Water and Water Environments*, Proceedings of the European seminar on water geography. London SOAS and Middlesex University, pp. 7–16.

Madani, I. M., Al-Shiryan, A., Lori, I., and Al-Khalifa, H. (1992) Public awareness and attitudes toward various uses of renovated water, *Environment International*, (18): 489–95.

Mahdi, Kamil (1996) Gulf agricultural policies and water scarcity, Contribution to the Symposium on Water and Arab Gulf Development: Problems and Policies, University of Exeter.

Major, David C. et al. (1996) Mexico City: Metaphor for the world's urban future, *Environment*, 38 (1): 32.

Mallat, Chibli (1995) The quest for water use principles: reflections on Shari'a and custom in the Middle East. In: Allan, J. A. et al. (eds), *Water in the Middle East: Legal, Political and Commercial Implications*, London: I. B. Tauris, pp. 127–37.

Mallat, Hyam (1995) Water laws in Lebanon. In: Allan, J. A. et al. (eds), *Water in the Middle East: Legal, Political and Commercial Implications*, London: I. B. Tauris, pp. 151–74.

Mandelbaum, Michael (1988) *The Fate of Nations: The Search for National Security in the Nineteenth and Twentieth Centuries*, Cambridge University Press.

Mastrull, Diane (1995) Flowing uphill, *The Economist*, (336): 36.

Mathews, Jessica Tuchman (1989) Redefining security, *Foreign Affairs*, **68** (2): 162–77.

Matthew, Richard A. (1995) Environmental security: demystifying the concept, clarifying the stakes. In: *Woodrow Wilson Center's Environmental Change and Security Project Report*, Issue 1.

McCaffrey, S. C. (1992) Background and overview of the International Law Commission's study of the non-navigational uses of international watercourses, *Colorado Journal of International Environmental Law and Policy*, 3 (1): 17–29.

McCaffrey, S. C. (1993) Water, politics, and international law. In Gleick, P. H. (ed.), *Water in Crisis: A Guide to the World's Fresh Water Resources*, Oxford University Press, pp. 92–104.

McCormick, J. (1989) *The Global Environmental Movement: Reclaiming Paradise*, London: Belhaven.

McDonald, Adrian and Kay, David (1988) *Water Resources: Issues and Strategies*, London and New York: Longman.

McLachlan, K. S. and Beaumont, P. (eds) (1985) *Agricultural Development in the Middle East*, New York: John Wiley.

Meadows, D. H., Meadows, D. L., Randers, J., and Behrens III, W. (1983) *The Limits to Growth*, London: Pan.

MEED (1995) 'Water–Special Report', *Middle East Economic Digest*, 27 January: 8–13.

Mehta, J. S. (1986) The Indus Water Treaty: a case study in the resolution of international river basin conflict. In: Vlachos, E. et al. (eds), *The Management of International River Basin Conflicts*, Washington, DC: George Washington University, ch. 2.1: 24pp.

Meybeck, Michael, Chapman, Deborah and Helmer, Richard (1989) *Global Freshwater Quality: A First Assessment*, Oxford: Blackwell.

Micklin, P. P. (1989) *The Water Management Crisis in Soviet Central Asia*, Final report to the National Council for Soviet and East European Research, Washington, DC. February.

Midgley, Mary (1983) *Animals and Why They Matter*, Harmondsworth: Penguin.

Miller, David (1984) *Anarchism*, London and Melbourne: J. M. Dent.

Milton, Kay (ed.) (1993) *Environmentalism: The View from Anthropology*, London and New York: Routledge.

Mooradian, Moorad (1992) Wars over water: population increase and dwindling water resources threaten the stability of the Middle East, *AIM: Armenian International Magazine*, http://www2.elibrary.com.

Mori, S. (1987) Water resources and development, *Archiv fur Hydrobiologie*, (28): 1–7, Shiga University, Hikone, Japan.

Mumme, Stephen P. (1995) New challenges for US–Mexico water resources management. In: Blake, Gerald H. et al. (eds), *The Peaceful Management of Transboundary Resources*, London: Graham & Trotman, pp. 261–76.

Munasinghe, Mohan (1990) Managing water resources to avoid environmental degradation: policy analysis and application, Environment Working paper no. 41, Washington, DC: World Bank.

Murakami, Masahiro (1995) *Managing Water for Peace in the Middle East: Alternative Strategies*, Tokyo: United Nations University Press.

Murakami, Masahiro and Musiake, Katsumi (1994) The Jordan River and the Litani. In: Biswas, Asit K. (ed.), *International Waters of the Middle East: From Euphrates–Tigris to Nile*, Oxford University Press, pp. 117–55.

Murakami, Masahiro et al. (1995) Technopolitical alternative strategies in interstate regional development of the Jordan Rift Valley beyond the peace, *Water International*, 20 (4): 188–96.

Murphy, Irene L. and Sabadell, J. Eleonora (1986) International river basins: a policy model for conflict resolution, *Resource Policy*, 12 (1): 133–44.

Musallam, Ramzi (1990) *Water: Source of Conflict in the Middle East in the 1990s*, London: Gulf Centre for Strategic Studies.

Musschenga, Bert (1994) Liberal neutrality and the justification of environmental conservation. In: Zweers, Wim and Boersema, Jan J. (eds), *Ecology, Technology and Culture*, Cambridge: White Horse Press, pp. 164–74.

Mustafa, I. (1994) The Arab–Israeli conflict over water resources. In: Isaac, J. and Shuval, H. (eds) *Water and Peace in the Middle East*, Amsterdam and London: Elsevier. pp. 123–34.

Myers, Norman (1986) The environmental dimension to security issues, *The Environmentalist*, 6 (4): 251–7.

Myers, Norman (1989) Environment and security, *Foreign Policy* (74): 23–41.

Myers, Norman (1993) *Ultimate Security: The Environmental Basis of Political Stability*, New York and London: Norton.

Naess, A. (1984) Intuition, intrinsic value and deep ecology, *The Ecologist*, 14 (5/6): 201–203.

Naess, A. (1989) *Ecology, Community and Lifestyle*, Cambridge University Press.

Naff, Thomas (1991) *Water Issue in Iraq*, Philadelphia: Associates for Middle East Research.

Naff, Thomas (1993) Water: that peculiar substance, *Research and Exploration*, 9 (Special issue): 7–17.

Naff, Thomas (1994a) A case for demand-side water management. In: Isaac, J. and Shuval, H. (eds), *Water and Peace in the Middle East*, Amsterdam and London: Elsevier, pp. 83–92.

Naff, Thomas (1994b) Conflict and water use in the Middle East. In: Rogers, P. and Lydon, P. (eds), *Water in the Arab World: Perspectives and Progress*, Cambridge, Mass.: Harvard University Press, pp. 253–84.

Naff, Thomas and Matson, Ruth C. (eds) (1984) *Water in the Middle East: Conflict or Cooperation?*, Boulder, Colo.: Westview.

Narveson, Jan (1997) Resources, environmental concerns and liberty. A paper presented at the University of Newcastle Upon Tyne on 2 May.

Nash, Linda (1993) Water quality and health. In: Gleick, Peter H. (ed.) *Water in Crisis: A Guide to the World's Fresh Water Resources*. Oxford University Press, pp. 25–39.

Ngan, Lai Ling Elizabeth (1991) *Water in Ancient Israelite Society During the Period of the Monarchy: Substance and Symbol*, PhD Thesis, Golden Gate Baptist Theological Seminary.

Nicholson-Lord, David (1995) UK still one of main polluters of North Sea, *Independent*, 3 June, p. 8.
Nijim, B. K. (1990) Water resources in history of the Palestinian conflict, *GeoJournal*, 21 (4): 317–23.
O'Riordan, T. (1981) *Environmentalism*, 2nd edn, London: Pion.
Omicinski, John (1996) Is the Earth running out of water? Gannett News Service, 09-12-1996, http://www2.elibrary.com.
Oppenheim, L. (1948) *International Law*, London: Longman.
Ostrom, Elinor (1990) *Governing the Commons: The Evolution of Institutions for Collective Action*, Cambridge University Press.
Overman, Michael (1976) *Water: Solutions to a Problem of Supply and Demand*, London: Open University Press.
Ozanne, Julian (1996) Winds of change sweep the Golan Heights, *Financial Times*, 7 February: 4.
Paehlke, Robert C. (1989) *Environmentalism and the Future of Progressive Politics*, New Haven, Conn.: Yale University Press.
Painton, Frederick (1990) Where the sky stays dark: the lifting of the Iron Curtain reveals the planet's most polluted region, *TIME*, 28 May: 40.
Papp, Daniel S. (1991) *Contemporary International Relations*, New York: Macmillan.
Pearce, Fred (1991) Wells of the conflict in the Middle East, *New Scientist*, June: 36–40.
Pearce, Fred and Hudson, David (1991) Rivers of blood, waters of hope, *Guardian*, 6 December: 29.
Pepper, D. (1984) *The Roots of Modern Environmentalism*, Beckenham: Croom Helm.
Perera, Judith (1981) Water Politics, *The Middle East*, (February): 47–54.
Peres, Shimon with Arye Naor (1993) *The New Middle East*, Shaftesbury: Element.
Peters, Joel (1994) *Building Peace in the Middle East: The Multilateral Arab–Israeli Peace Talks*, Middle East Programme Report, Royal Institute of International Affairs, London.
Pirages, D. (ed.) (1977) *The Sustainable Society*, New York: Praeger.
Pirages, D. (1991) Environmental security and social evolution, *International Studies Notes*, 16 (1): 8–13.
Pisani, Edgard (1995) The management of water as an essential and rare commodity, *Water International*, 20 (1): 29–31.
Plusquellec, Hervé et al. (1994) *Modern Water Control in Irrigation: Concepts, Issues, and Applications*, World Bank Technical Paper no. 246, Washington, DC: World Bank.
Pope, Hugh (1995) Ataturk Dam: a source of power and of conflict, *Los Angeles Times (Home Edition, 5–30–1995)*: 4. http://www2.elibrary.com.
Porritt, J. (1986) *Seeing Green*, Oxford: Blackwell.
Porritt, J. and Winner, D. (1988) *The Coming of the Greens*, London: Fontana.
Postel, Sandra (1989) *Water for Agriculture: Facing the Limits*, Worldwatch Paper 93, Washington, DC.
Postel, Sandra (1992) *The Last Oasis: Facing Water Scarcity*, London: Earthscan.
Pringle, Laurence (1982) *Water: The Next Great Resource Battle*, London and New York: Macmillan.
Quingquan, Shi (1994) Environment and resettlement. In: Sun, Peter (ed.),

Multipurpose River Basin Development in China, Washington, DC: The World Bank, pp. 63–5.

Radosovich, G. E. (1979) Western water law, *Water Spectrum*, **11** (3): 1–9.

Raphael, C. Nicholas and Shaibi, Hussain T. (1984) Water Resources for At Taif, Saudi Arabia: a study of alternative sources for an expanding urban area, *Geographical Journal*, **150** (2): 183–91.

Redclift, Michael and Benton, Ted (eds) (1994) *Social Theory and the Global Environment*, London: Routledge.

Reisner, Marc (1990) *Cadillac Desert: The American West and Its Disappearing Water*, London: Secker & Warburg.

Renner, Michael G. (1989a) National security: the economic and environmental dimensions, Worldwatch paper no. 89, Washington DC: Worldwatch Institute.

Renner, Michael G. (1989b) Forging Environmental Alliances, *World Watch*, (Nov.–Dec.): 8–15.

Renner, Michael G. (1989c) Enhancing global security. In: Brown, L. R. et al. (eds), *State of the World 1989*, New York: Norton, pp. 132–53.

Repetto, Robert, (1986) *Skimming the Water: Rent-seeking and the Performance of Public Irrigation Systems*, Washington, DC: World Resources Institute.

Richards, Charles (1992) 'Sink or swim' warning to Mid-East on water: Agreement must be reached on a precious resource if disputes are not to spill over into war, *Independent*, 30 December: 9.

Roberts, Neil (1991) Geopolitics and the Euphrates' water resources, *Geography*, **76** (Apr.): 157–9.

Rogers, Peter and Lydon, P. (1994) *Water in the Arab World: Perspectives and Prognoses*, Cambridge, Mass.: Harvard University Press.

Rowlands, Ian H. (1991) The security challenges of global environmental change, *Washington Quarterly*, **14** (1): 99–114.

Rzoska, J. (1980) Euphrates and Tigris Mesopotamian Ecology and Destiny, The Hague: W. Junk.

Saif, Abdul-Aziz M. (1987) *Optimization of Scarce Water Resources for Irrigation in Yemen*. PhD Thesis, University of Southampton, UK.

Sancton, Thomas A. (1989) What on earth are we doing?, *TIME*, 2 January: 24.

Schiffler, Manuel (1995) Sustainable development of water resources in Jordan. In: Allan, J. A. et al. (eds) *Water in the Middle East: Legal, Political and Commercial Implications*, London and New York: I. B. Tauris, pp. 239–59.

Schmida, L. (1984) Israel's drive for water, *The Link*, **17** (4): 1–3.

Schoon, Nicholas (1995) UN summit fuels global warning debate, *Independent*, 29 March: 12.

Schrijver, Nico (1989) International organization for environmental security, *Bulletin of Peace Proposals*, **20** (2): 115–22.

Schulte-Wulwer-Leidig, A. (1992) International Commission for the Protection of the Rhine Against Pollution – the integrated ecosystem approach for the Rhine, *European Water Pollution Control*, **2** (3): 37–41.

Science Council of Canada (1988) *Water 2020: Sustainable Use for Water in the 21st century*, Report no. 40.

Serageldin, Ismail (1994) *Water Supply, Sanitation, and Environmental Sustainability*, Washington, DC: World Bank.

Serageldin, Ismail (1995) Water resources management: a new policy for sustainable future, *Water International*, **20** (1): 15–21.

Serageldin, Ismail and Steer, Andrew (eds) (1994) *Valuing the Environment*, Washington, DC: The World Bank.
Shabad, Theodore (1986) News Notes: Soviet Decree officially cancels north–south water transfer projects, *Soviet Geography*, 27 (8): 601–3.
Shadid, Anthony (1995) Cairo may see waters of Nile diminish, *Los Angeles Times*, (Bulldog Edition, 12–17–1995, pp. A-38) http://www2.elibrary.com.
Shahin, Mamdouh (1989) Review and assessment of water resources in the Arab region, *Water International*, 14 (4): 206–19.
Shapland, Greg (1995) Policy options for downstream states in the Middle East. In: Allan, J. A. et al. (eds) *Water in the Middle East: Legal, Political and Commercial Implications*, London and New York: I. B. Tauris, pp. 301–23.
Shapland, Greg (1997) *Rivers of Discord: International Water Disputes in the Middle East*, London: Hurst.
Shuval, Hillel I. (1992) Approaches to resolving the water conflicts between Israel and her neighbors – a regional water-for-peace plan, *Water International*, 17 (3): 133–43.
Smith, Charles D. (1993) Arabs. In: *Grolier Multimedia Encyclopedia*, Release 6.
Smith, Rodney T. (1988) *Trading Water: An Economic and Legal Framework for Water Marketing*, Washington, DC: The Council of State Policy and Planning Agencies.
Solanes, Miguel (1992) Legal and institutional aspects of river basin development, *Water International*, 17 (3): 116–23.
Sorensen, T. C. (1990) Rethinking national security, *Foreign Affairs*, 69 (3): 1–18.
Soroos, M. S. (1986) *Beyond Sovereignty: the Challenge of Global Policy*, Columbia: University of South Carolina Press.
Specter, Michael (1995) The city called 'Bukhara', *New York Times*, 19 April.
Spooner, Brian (1993) Bedouin. In: *The New Grolier Multimedia Encyclopedia*, Release 6, Grolier Inc, USA.
Spretnak, C. and Capra, F. (1985) *Green Politics: The Global Promise*, London: Paladin.
Starr, Joyce R. (1991) Water Wars, *Foreign Policy*, (82): 17–36.
Starr, Joyce R. (1993) Quest for water from the Biblical times to the present, *Environmental Science and Technology*, 27 (7): 1264–67.
Starr, Joyce R. (1995) *Covenant over Middle Eastern Waters: Key to World Survival*, New York: Henry Holt.
Starr, Joyce R. and Stoll, Daniel C. (1987) *U.S. Foreign Policy on Water Resources in the Middle East*, Washington, DC: The Center for Strategic and International Studies.
Starr, Joyce R. and Stoll, Daniel C. (1988) *The Politics of Scarcity: Water in the Middle East*, Boulder, Colo.: Westview.
Stumm, Werner (1986) Water, an endangered ecosystem, *AMBIO*, 15 (7): 201–7.
Suliman, Mohamed (1992) *Civil War in Sudan: The Impact of Ecological Degradation*. ENCOP Report no. 1, Centre for Security Studies and Conflict Research, Swiss Federal Institute of Technology, Zurich.
Sun, Peter (1994) *Multipurpose River Basin Development in China*, Washington, DC: The World Bank.
Sylvan, Richard and Bennett, David (1994) *The Greening of Ethics*, Cambridge: White Horse Press.

Teclaff, Ludwik (1967) *The River Basin in History and in Law*, New York: United Nations Press.
Teclaff, Ludwik (1976) Harmonising water resource development and use with environmental protection in municipal and international law, *Natural Resource Journal*, **16**: 807–58.
Teclaff, Ludwik (1977) *Legal and Institutional Responses to Growing Water Demand*. FAO Legislative Study no. 14, Rome: FAO.
Teclaff, Ludwik (1985) *Water Law in Historical Perspective*, Buffalo and New York: Hein.
Teclaff, Ludwik and Utton, A. (1981) *International Groundwater Law*, London: Oceana.
Tekeli, Sahim (1990) Turkey seeks reconciliation for the water issue induced by the Southeastern Anatolia Project (GAP), *Water International*, **15** (4): 206–16.
Temperley, Tom G. (1992) Saudi water planners look out to sea, *Water and Environment International*, **2** (14): 18–27.
Thomas, Caroline (1992) *The Environment in International Relations*, London: RIIA.
Thomas, Caroline and Howlett, Darryl (eds) (1993) *Resource Politics: Freshwater and Regional Relations*, Buckingham: Open University Press.
Thompson, Roy L. (1978) Water as a source of conflict, *Strategic Review*, (6): 62–71.
Tickner, A. (1992) *Gender in International Relations: Feminist Perspectives on Achieving Global Security*, New York: Columbia University Press.
Timberlake, Lloyd (1984) The emergence of environment awareness in the West. In: Sardar, Ziauddin (ed.) *The Touch of Midas – Science, Values and Environment in Islam and the West*, Manchester University Press, pp. 123–33.
Timberlake, Lloyd. and Tinker, J. (1984) *Environment and Conflict*, Earthscan Briefing Document no. 40, London.
Timberlake, Lloyd. and Tinker, J. (1985) The environmental origins of political conflict. *Socialist Review*, **15** (6): 57–75.
Tokar, B. (1987) *The Green Alternative*, San Pedro: R. and E. Miles.
Tomanbay, Mehmet (1993) Sharing the Euphrates: Turkey, *Research and Exploration*, **9** (Special issue): 53–61.
Trager, James (1994) *The People's Chronology*. CD Rom version, Henry Holt & Co.
Tuijl, Willem Van (1993) *Improving Water Use in Agriculture. Experiences in the Middle East and North Africa*, World Bank Technical Paper no. 201, Washington, DC: The World Bank.
Ullman, Richard H. (1983) Redefining security, *International Security*, **8** (1): 129–53.
UN (1978) *United Nations Register of International Rivers*, Oxford: Pergamon.
UN (1983) *The Law of the Sea – Official Text of the United Nations Convention on the Law of the Sea with Annexes and Index*, New York: United Nations.
UN (1984) *Treaties Concerning the Utilisation of International Water Courses for Other Purposes than Navigation*, Natural Resources/Water Series, no. 13, New York: UN Department of Technical Cooperation for Development.
UN (1997) *Convention on the Law of the Non-navigational Uses of International Watercourses*, Distr. GENERAL A/RES/51/229, UN General Assembly

Fifty-first session (Agenda item 144), Resolution adopted by the General Assembly in 99th plenary meeting on 21 May 1997.
UNCED (1992) Report of the United Nations Conference on Environment and Development (Agenda 21). (ch. 18) doc. no. A/CONF.151/26 (vol. II).
UNGA (1970) *Progressive Development and Codification of the Rules of International Law Relating to International Watercourses*. UNGA Resolution no. 2669 (XXV) of 8 December 1970, New York.
UNWC (1977) *Report of the United Nations Water Conference*, Mar del Plata, 14–25 March 1977, UN Document No. E/CONF.70/29, New York.
UNWC (1978) *Water Management and Development*. Proceedings of the United Nations Water Conference, Mar del Plata, Argentina March 1977, New York: Pergamon.
US Army Corps of Engineers (1991) *Water in the Sand: A Survey of Middle East Water Issues*, Washington, DC: US Army (prepared by Robertson, W.; Priscoli, J.; Brumbaugh, R.)
Van Dijk, Paul (1994) Theological–Anthropological reflections on the environmental issue. In: Zweers, Wim and Boersema, Jan J. (eds), *Ecology, Technology and Culture*, Cambridge: White Horse Press, pp. 56–62.
Vermeersch, Etienne (1994) The future of environmental philosophy. In: Zweers, Wim and Boersema, Jan J. (eds), *Ecology, Technology and Culture*, Cambridge: White Horse Press, pp. 272–86.
Viessman Jr., W. (1990) A framework for reshaping water management, *Environment*, (32): 10–11.
Vlachos, E., Webb, A. C. and Murphy, I. L. (1986) *The Management of International River Basin Conflicts*, Washington, DC: George Washington University.
Vlachos, Evan (1990) Prologue: water, peace and conflict management, *Water International*, 15 (4): 185–8.
Vogler, John (1993) Security and global environmental change, *Conflict Processes*, 1 (2): 1–13.
Vogler, John and Imber, Mark F. (eds) (1996) *The Environment and International Relations*, London and New York: Routledge.
Wagnick, K. (1990) *Worldwide Desalting Plants Inventory*, Report no. 11, Wagnick Consulting, Gnarrenburg, International Desalination Association.
Wagnick, K. (1992) *World Wide Desalting Plants Inventory*, Report No. 12, Wagnick Consulting, Gnassenburg International Desalination Association.
Wakil, Mikhail (1993) Sharing the Euphrates: Syria, *Research and Exploration*, 9 (Special issue): 63–71.
Warburg, Philip (1995) *Middle East Environmental Co-operation*, Policy Briefs no. 04, Institute on Global Conflict and Co-operation, San Diego.
Ward, Christopher (1997) *Yemen: Towards a Water Strategy*, Washington, DC: World Bank.
Ward, Colin (1997) *Reflected in Water: A Crisis of Social Responsibility*, London: Cassell.
Warford, Jeremy J. (1968) Water supply. In: Turvey, R. (ed.), *Public Enterprise*, Harmondsworth: Penguin.
Warner, Frederick (1991) The environmental consequences of the Gulf War, *Environment*, 33 (5): 7–9, 25–26.
Waterbury, John (1994) Transboundary water and the challenge of international co-operation in the Middle East. In: Rogers, P. and Lydon,

P. (eds), *Water in the Arab World: Perspectives and Progress*, Cambridge, Mass.: Harvard University Press, pp. 39–64.
Watson, A. (1983) *Agricultural Innovation in the Early Islamic World*, Cambridge University Press.
Weber, P. (1991) Desalination appeal evaporates, *World Watch*, 4 (6): 8–9.
WE (1992) Trans-Gulf pipeline talks, *Water and Environment*, 2 (14): 6.
Weisgall, M. W. (ed.) (1977) *The Letters and Papers of Chaim Weizman*, Jerusalem: Israel University Press.
WEI (1992) Saudi-Arabia moves to expand desalt capacity, *Water and Environment International*, 2 (19:2).
Weitz, Joseph (1965) *My Diary and Letters to the Children*, vol. 4, Massada, Tel-Aviv (in Hebrew).
Westing, Arthur H. (ed.) (1986) *Global Resources and International Conflict: Environmental Factors in Strategic Policy and Action*, Oxford University Press.
Willetts, Peter (1996) Who cares about the environment? In: Vogler, John and Imber, Mark F. (eds), *The Environment and International Relations*, London and New York: Routledge, pp. 120–37.
Williams, Marc (1996) International political economy and global environmental change. In: Vogler, John and Imber, Mark F. (eds), *The Environment and International Relations*, London and New York: Routledge, pp. 41–58.
Winpenny, James (1991) *Values for the Environment: A Guide to Economic Appraisal*, London: HMSO.
Winpenny, James (1992) Powerless and thirsty? Prospects for energy and water in developing countries, *Utilities Policy*, Special Edition (Oct): 290–5.
Winpenny, James (1994) *Managing Water as an Economic Resource*, London and New York: Routledge.
Wittfogel, Karl (1956) *Oriental Despotism: A Comparative Study of Total Power*, New Haven, Conn.: Yale University Press.
Wolf, Aaron T. (1995) *Hydropolitics along the Jordan River: Scarce Water and its Impact on the Arab-Israeli Conflict*, Tokyo, United Nations University Press.
Wolf, Aaron T. (1997) Conflict and Co-operation along International Waterways, a paper presented at the ADC Millennium meeting on *International Water Management in the Twenty First Century*, Valencia, Spain, 18–20 December.
Wolf, Aaron T. and Dinar, Ariel (1994) Middle East Hydropolitics and Equity Measures for Water Sharing Agreements, *Journal of Social, Political and Economic Studies* (19): 69–93.
Wollman, Stuart (1996) Environmental hydrology activities in Israel. IAEH Activity Report. Hydro Web Homepage (URL:http://www.hydroweb.com/).
World Bank (1990) *Water Resources Management: A Policy Paper*, Washington, DC: The World Bank.
World Bank (1992a) *World Development Report 1992 – Development and the Environment*, New York: Oxford University Press.
World Bank (1992b) *Water Supply and Sanitation Projects: The Bank's Experience 1967–89*, Washington, DC.
World Bank (1993) *Water resources management*, A World Bank Policy Study, Washington, DC: World Bank.
World Bank (1994) *A Strategy for Managing Water in the Middle East and North Africa*, Washington, DC: World Bank.

World Bank (1995) *From Scarcity to Security: Averting a Water Crisis in the Middle East and North Africa*, Washington, DC: World Bank.
World Resources Institute (1990) *World Resources 1990–91*, Oxford University Press.
World Resources Institute (1996/7) *World Resources 1996–97*, World Resources Institute, Washington, DC. Electronic version: http://www.wri.org/wri/wr-96-97/96tocful.html.
World Water (1984) Israel accused of stealing water, *World Water*, 7 (11): 7.
Wu, Yi Ao (1994) General review, experience, and problems. In: Sun, Peter (ed.), *Multipurpose River Basin Development in China*, Washington, DC: World Bank, pp. 17–22.
Yergin, Daniel (1991) *The Prize: The Epic Quest for Oil, Money, and Power*, New York: Simon & Schuster.
Young, Robert A. (1996) *Measuring Economic Benefits for Water Investments and Policies*, World Bank Technical Paper no. 338, Washington, DC: World Bank.
Yunus Khan, M. (1990) Boundary water conflict between India and Pakistan, *Water International*, 15 (4): 195–99.
Zaman, Munir (1983) Ganges basin development: a long-term problem and some short-term options. In: Zaman, M. et al. (eds), *River Basin Development*, Dublin: Tycooly, pp. 99–109.
Zarour, Hisham and Isaac, Jad (1993) Nature's apportionment and the open market: a promising solution to the Arab–Israeli water conflict, *Water International*, 18 (1): 40–53.
Zarour, Hisham and Isaac, Jad (1994) A novel approach to the allocation of international water resources. In: Isaac, J. and Shuval, H. (eds), *Water and Peace in the Middle East*, Amsterdam and London: Elsevier, pp. 389–98.
Zeitouni, Naomi et al. (1994) Water sharing through trade in markets for water rights: an illustrative application to the Middle East. In: Isaac, J. and Shuval, H. (eds), *Water and Peace in the Middle East*, Amsterdam and London: Elsevier, pp. 399–413.

Index

Note: the index does not include material contained in the maps, tables, figures or chapter end-notes.

Abbasid 130
Abdulrazzak, M. 167, 181, 182, 188, 189, 190
Abi-Eshuh 127
Abimelech 3
Abraham 3
Abramovitz, J. N. 6
Abu Dhabi Solar Desalination Plant 191
Abu Dibis 135
Achilles' heel 105
Aczel, J. 69
Adam 2
Adams, R. McC. 124
Africa 63, 67
  North 61, 171, 177, 179, 211
  Northern 168
African 41
  Development Bank (ABD) 153
  North 45
Agenda 21 7, 30, 56
Agnew, C. 175, 176, 179
Ahmad, M. 188, 198, 200
Al-Alawi, J. 167, 182, 188, 189, 190
Alam, U. Z. xi, 9, 20, 37, 107
Alaska 71
Aleppo 133, 134
Alexander 127
Alexandretta (Hatay province) 137
Alexandria 60
Algeria 177
Al-Kuwari, A. K. 183
Allah 3
Allan, J. A. 19, 25, 33, 80, 110, 168, 179, 181, 182, 184, 192, 198
al-Qurna 122
Al-Sheikh, H. M. H. 188
Al-Zubari, W. K. 167, 192, 194
America 69
  Far West 70
  North 13, 60, 67, 80
  Northeast 70
  Southwest 67, 71
  West 70
America's ecology 69
American 72, 211
  intervention 116
  native people 211
  officials 116
Americans 24, 69, 71
  North 71
Amir, D. 217
Amritsar 76
Amu Darya River 76, 77, 79
Anatolia 156
Anatolian Plateau 119
Anderson, E. W. 121, 164, 170, 175, 184, 196, 201
Andhra Pradesh 76
Anglo-American 213
Ankara 133, 148, 155, 195
an-Nafud 167
Antarctic 66, 189, 198
AOAD (Arab Organisation for Agricultural Development) 187
Arab 5, 92, 98, 99, 103, 105, 106, 110, 111, 112, 142, 143, 147, 148, 149, 153, 171, 173, 178, 179, 191, 196, 213

Index

countries 153
Empire 130
Iskenderun 149
League 95, 153
nationalism 143
Arabia 126, 167, 171, 172, 175, 178, 179
  Ancient 171
  Felix 167, 180
  Modern 178, 181
  Northern 171
Arabian Peninsula 11, 13, 41, 118, 124, 156, 161, 164–202, 210, 211, 213, 214
Arabic 1, 5, 85, 171
Arab–Islamic era 130
Arab–Israeli 113
  conflict 81, 86, 87, 93, 98, 112, 113, 215
  War (1967)19; (1973) 30; (1948) 95
Arafat, Y. 106
Aral Sea 28, 76, 77, 78, 79
Aralsk 77
Aramean Kingdom 125
Arar, A. 192
Argentina 7, 207
Arizona 70, 71, 72
Armitage, R. 116
Asi (Orontes) River 149, 153
Asia 72
  Central 13, 24, 28, 41, 61, 67, 76–9
  Southeast 80
Asian 2
  Republics, Central 142
Asir Mountains 167, 168
Assurbanipal, King of Assyria 127, 175
Assyrian 126, 171
  Empire 94
  Kingdom 125
Assyrians 94
Aswan (High) Dam 19, 41, 137
Ataturk Dam 145, 146, 154
Athens 60
Attfield, R. 52, 219

Ba'ath party 143
Ba'athi rulers 137
Babbitt, B. 70
Babylon 60, 125, 126, 133
  Hanging Gardens of 2
Babylonian Kingdom 125
Babylonians 41, 94
Baghdad 158, 191, 201
Bahrain 168, 180, 181, 182, 187, 190, 194, 195, 200, 201, 217
Bailey, B. 35, 207
Bakour, Y. 124, 133, 147, 160
Balikh River 122
Baltic Sea 69
Banias River 89, 95
Banks, P. A. 192, 193
Barham, J. 116, 122, 154, 155
Basra 158
Batinah 187
Beaumont, P. 107, 122, 132
Bedouin 99, 171, 172, 174, 175, 179
Beijing 29, 74
Benares 2
Ben-Gurion, D. 100, 102
Ben-Shahar, H. 110
Berkes, F. 211
Beschorner, N. 8, 91, 121, 122, 123
Bible 2, 97
Bilen, O. 141, 150
Bingcai, W. 74
Biswas, A. K. 20, 21, 60, 62
Black Sea 68, 119
Blake G. H. 7, 35
Blom, G. 35
Bodgener, J. 160, 177, 187, 201
Boersema, J. J. 56, 218
Boston University 41
Botswana 63
Britain 53, 68
British 94, 132
  government 95
  Mandate 93, 94, 95, 98, 130, 133
  troops 125
Brown & Root 195

Buddha 2
Buddhism 2
Bukhara 78
Bulgaria 63
Bulloch, J. 9, 30, 86, 105, 106, 116, 122, 148, 150, 153, 154, 155, 156, 168, 172, 174, 176, 177, 184, 185, 195, 201, 202, 207
Bunyard, P. 48, 51, 219
Buraimi-Al ayn 187
Burdon, D. J. 168
Bushehr 195
Butler, D. 9, 21, 63, 106, 186
Buzan, B. 55

Cairo 41, 137
California 33, 70, 71
 Aqueduct 41, 70
 Central Valley 70
 Southern 70
 Water Bank 33
Calvert, P. 72
Cambodia 63
Canada 71
 Western 71
Canadian 217
Cano, G. 19
Caponera, D. A. 173
Carthage 60
Central Intelligence Agency (CIA) 116
Centre for Strategic and International Studies 8, 106
Ceyhan River 195
Chalabi, H. 133, 134, 135, 137, 148, 149
Charnock, A. 191
Cheerapunji region 43
Chesnoff, R. Z. 161
China 2, 13, 41, 61, 66, 67, 72, 73, 74
 Environmental Protection Law 73
 Ministry of Water Resources 73
 Northern 67
 North China Plain 74

 State Environmental Protection Committee 73
Chinese government 73
Chomchai, P. 19
Christian regime 107
Christianity 2
Ciller, T. 195
Clarke, R. 8, 22, 33, 34, 43, 64, 65, 66, 67, 76, 77, 78, 80, 82
Cold War 55, 111, 157, 214
Colorado 71
 River 24, 72
Congo, the 63
Congress of Ministers of Agriculture 188
Convention on the Protection and Use of Transboundary Watercourses and International Lakes 138–9
Cooley, J. K. 86, 95, 105, 106, 116, 207
Cotgrove, S. 61
Crusades 94
Cyrus 126, 127

Dabelko, D. D. 46
Dabelko, G. D. 46
Dalby, S. 218
Damascus 155
Dan River 89, 95
Danube, River 68
Darwish, A. 9, 30, 86, 106, 116, 122, 148, 150, 153, 155, 156, 168, 172, 174, 176, 177, 184, 185, 195, 201, 202, 207
Davey, T. 27, 62
de Jong, R. L. 168, 183, 188, 195
Dead Sea 85, 89, 95, 107
Delhi 29
Dellapenna, J. 9, 20, 39, 113, 202
Demirel, S. 147, 154
Department of Irrigation 133
de-Shalit, A. 96, 97, 99, 100, 101, 102, 110
deShazo, R. 106, 119

Index 247

Deudney, D. xi, 9, 20, 218
Dillman, J. 107
Dinar, A. 36, 37
Dobson, A. 48, 49, 50, 52
Dolatyar, M. 47, 81, 104, 219
Drower 126, 175
Duma, C. 119
Duwais, A.-A. M. 186
Dyer, H. C. 17, 55, 218

Earth Observation Satellite Company (EOSAT) 142
Earth 16
East–West 117, 118, 137, 138, 160, 213, 214, 215
Eaton, D. J. 94, 95, 97, 98, 104
Eaton, J. W. 94, 95, 97, 98, 104
Eban, E. 215
Eckersley, R. 51
ECWA (Economic Commission for Western Asia) 183
Eden, Garden of 2, 80
  Paradise of 2
Egypt 19, 30, 41, 63, 185
Egyptian 137, 148
  Sahara 41
Elam 126
El-Baz, F. 41
Elhance, A. 9, 20
Elliott, M. 62, 219
Elmer-Dewitt, P. 41, 71
El Morr, A. 150
El-Nashar, A. M. 191
Encarta 77, 167, 171
Engelmann, K. E. 76, 77
Environmental
  Code of Conduct 217
  Issues Foundation 157
Esarhaddon 127
Euphrates River 30, 85, 90, 119, 120, 121, 122, 123, 125, 126, 130, 132, 133, 134, 135, 136, 137, 138, 139, 140, 141, 144, 145, 147, 148, 149, 150, 152, 154, 155, 156, 158, 160, 161, 214

Euphrates–Tigris basin/ valley 4, 11, 13, 80, 114, 116–161, 168, 195, 213, 214
Europe 13, 60, 63, 64, 67, 68, 69, 96
  Central 68
  Eastern 69
  Western 68
European
  Coal and Steel Community 213
  Investment Bank (EIB) 153
  Union (EU) 217
Europeans 68
Eve 2

Fahd, King 199
Fahim, H. M. 179, 199, 202
Falkenmark, M. 6, 8, 15, 18, 21, 37, 38, 60, 62, 64, 67, 82, 123, 156, 207
FAO (Food and Agriculture Organization) 183, 188
Farnham, H. 36
Feitelson, E. 92, 97, 103, 104, 108, 109, 110, 111
Financial Times 8, 22, 154
Fox, I. K. 37
France 68, 134, 194
Franco–British Convention (1920) 133
Franco–Turkish agreement (1921) 133
Frankel, N. 147, 148, 152
French
  government 95, 133
  Mandate 94, 130, 133
  Turkish Protocol (1930) 134
Frey, F. W. 35, 36, 37, 38
Friendship and Neighbourly Relations Convention (1926) 133
Friendship and Good Neighbourly Relations Treaty (1946) 134, 140, 143, 147

Galilee
  Northern 95

Galilee – *continued*
  Galilee, Sea of 89, 98, 100, 108
Gambia 63
Gandhi, I. 76
Ganges River 2
GAP (*Güneydogu Anadolu Projesi*) 24, 41, 144, 145, 146, 152, 153, 154, 155, 157
Gare, A. 219
Gaza 81, 92
  Strip 92, 93
Genesis 2, 3
Geographical Information System (GIS) 142
Georgia 70
German 3
Germany 68
Gibbons, D. C. 27
Gischler, C. E. 90, 170, 188
Gleick, P. H. 9, 17, 19, 21, 29, 33, 35, 37, 38, 62, 63, 64, 78, 82, 86, 89, 117, 121, 122, 123, 125, 126, 127, 136, 175, 189, 192, 201, 202, 207, 210, 212, 216
Glen Canyon 72
Global Water Authority 35
God 2, 5, 171, 178, 211
Golan Heights 105
Goldberg, D. 153
Golubev, G. N. 76, 78
Gosh, H. 199
Gowers, A. 86, 136, 195, 197
Great Man-Made River 24, 41, 62
Great Plains 71
Great Recycling and Northern Development (GRAND) Canal 71
Great Salt Lake 71
Greeks 94
Grolier 75, 76, 77, 119, 121, 167
Grove-White, R. 48, 52
Gruen, G. E. 207
Gujarat 76
Gulf Centre for Strategic Studies 8

Gulf War *see under* Persian
Gunter, M. M. 154
Gup, T. 69
Gurr, T. R. 62

Habbaniyah, Lake 135
Haddad, M. 37, 103
Hammurabi, Code of 5, 125, 127, 159
Harayan 76
Hardan, A. 122, 125, 130, 133, 135, 158
Harmon
  Doctrine 36
  US Attorney-General 36
Hasbani River 89, 95
Hassan, S. 76
Hatami, H. 125, 126, 127, 207
Hatay province (Alexandretta) 137, 149, 154
Hebrew 3
Helsinki Rules on the Uses of the Waters on International Rivers 38
Herodotus 126
Hewedy, A. 95, 107, 108
Hezekiah 127
Hillel, D. 3, 9, 41, 94, 99, 113, 116, 120, 121, 122, 123, 135, 136, 139, 145, 170, 172, 173, 174, 176, 215
Hindiya Barrage 133
Hindu 2, 76
Hirshleifer, J. 23
Hit 123
Hitti, P. K. 171
Holdren, J. P. 22, 207
Homer-Dixon, T. F. 43, 44, 46, 92, 93
Hormuz, Strait of 119
Hotten, R. 68
Howell, P. P. 19
Howlett, D. 36, 56
Hudson, D. 107
Hullah, Lake 98, 100
Hurewitz, J. 95

Index

India 20, 43, 67, 74, 75, 76
Indian
  Ocean 171
  subcontinent 13, 61
Indochina 34, 207
Indus River 4, 37, 76
International Conference on Water and Environment (ICWE) 1992 7
International Desalination Association (IDA) 189
International Law Association (ILA) 38
International Law Commission (ILC) 38, 139
Internet 15
Iran 79, 119, 154, 177, 195
Iran–Iraq War (1980–8) 185, 191
Iranian 191, 195
Iraq 20, 24, 80, 116, 118, 120, 121, 123, 130, 131, 132, 133, 134, 135, 136, 137, 138, 140, 141, 143, 144, 145, 146, 147, 148, 149, 150, 152, 153, 154, 157, 158, 160, 161, 168, 191, 195, 203
  Board of Development 134
  Ministry of Agrarian Reform 134
  Ministry of Development 134
Iraqi 136, 149, 152, 158, 191, 202, 203
Iraqis 158, 191, 201
Irvine, S. 48
Isaac, J. 94, 105, 106, 110
Islam 2, 172, 175, 219
Islamic 5, 173
  culture 211
  jurisprudence 173
  law 160
  water law 172
Israel 20, 24, 41, 66, 81, 87, 89, 90, 91, 92, 96, 97, 98, 99, 100, 102, 103–111, 185, 212, 213, 215
  Foreign Ministry 111

Information Service 89, 92, 100, 103, 105
Jordan Peace Treaty (1994) 216
State Comptroller 92, 108, 109
Israeli 103, 104, 106, 110, 111
  government 92, 97, 100
  scholars 102
  sovereignty 95
Israeli–Palestinian 113
  conflict 11, 114, 212
  negotiations 106
Israelis 87, 111, 215
Israelites 2
Istanbul 212

James Bay 71
Japan 2, 194
Japanese 22
Jerusalem 97
Jewish 97, 100
  immigrants 93, 98, 99, 103
  National Fund (JNF) 101
  national homeland 94, 96
  population 98, 102
  rabbis 173
  settlements 96, 99, 102
  soil 99
  soul 99
  state 93, 98, 104, 112, 215
Jewry (world) 62
Jews 87, 94, 95, 96, 97, 98, 100, 104, 105, 113, 215
Jezreel Valley 98
Joffe, E. G. H. 186
Joffe, G. 116, 122, 136, 138, 164, 183
Johnson, T. P. 54
Joint Technical Committee for Regional Waters (1980) 140, 161
Jonglei Canal project 19
Jordan 20, 81, 83–114, 185, 195, 212, 213, 216
  East 95
  Lower [Jordan] River 89
  River 13, 83–114, 135, 214, 217

250 *Index*

Jordan – *continued*
 River basin 11, 83, 85–114, 116, 117, 118, 124, 160, 210, 214, 215
 Upper [Jordan] River 89, 90
 Valley 95
 Valley Authority 95
Juarez aquifer system 72
Judaism 2
Judeo-Christian 3

Kally, E. 92
Karakaya Dam 140, 145, 153
Karakum Canal 41, 77
Karnataka 75
Karun River 195
Katko, T. S. 26
Kay, D. 21
Keban Dam 122, 138, 140, 144, 145
Keen, M. 186
Kesterton Reservoir 70
Khabur River 122
Khatib, H. 188
Khiva 78
Khordagui, H. 168, 182, 187, 190, 192, 193, 194, 196, 197, 198, 199
Kirmani, S. S. 20, 76
Kliot, N. 8, 85, 91, 92, 93, 95, 102, 127, 131, 132, 134, 135, 136, 140, 141, 144, 145, 148, 149, 152, 160, 161
Kneese, A. V. 27
Knesset 103
Kolars, J. 117, 123, 124, 131, 133, 134, 141, 142, 144, 146, 147, 156, 157, 160, 189, 215
Koran 2, 3
Koranic 2
Koveik River 133, 134
Krishna, R. 35, 39, 153
Kurdish 137, 154, 155, 214
Kurdistan Workers' Party (PKK) 154, 155
Kuwait 80, 164, 168, 169, 181, 182, 185, 189, 190, 191, 192, 193, 194, 195, 201

Crisis (1990–1) 111, 158, 164
Institute of Scientific Research 191

Lacayo, R. 192
Lagash 127
Latin America 72
Lausanne, Treaty of (1923) 133
League of Nations 98
Lean, G. 19, 21, 67, 69, 71, 82
Lebanon 79, 85, 89, 90, 91, 95, 105, 107, 108, 147, 150, 217
 Southern 105, 107
LeMarquand, D. G. 19, 37
Le Moigne, G. 45, 73, 152
Lemonick, M. D. 192
Levant, the 156
Libiszewski, S. 81, 112, 113, 216, 217
Libya 24, 41, 62
Linden, E. 41, 43, 70, 72, 74, 75, 77
Lipschutz, R. D. 22, 34, 207
Litani River 95, 107, 108
Lonergan, S. 9
Los Angeles 70, 211
Louisiana, Southern 71
Lowdermilk 95
Lowi, M. R. 85, 89, 90, 94, 99, 102, 103, 120, 121, 123, 137, 156, 207
Lundqvist, J. 32

Madani, I. M. 194
Madras 74
Madrid 111
Maharashtra 76
Mahdi, K. 180, 181, 182
Majzoub, T. 133, 134, 135, 137, 148, 149
Mali 19
Mallat, C. 5, 25, 173, 174, 184
Malthusian 40
Mandate period 137

Index 251

Mandelbaum, M. 102
Manzur, I. 5
Mar del Plata 7, 207
Marshall Plan 213
Marxism 51, 218
Mastrull, D. 40, 86
Mathews, J. T. 54
Matson, R. C. 8, 20, 38, 89, 90, 92, 93, 95, 99, 106, 116, 121, 131, 135, 157, 212
Mauritania 19
McCaffrey, S. C. 150
McDonald, A. 21
Meadows, D. H. 49
Mecca 2
Mediterranean 95, 119, 147, 171
Medusa bags 189
Mehta, J. S. 19, 20, 37
Memphis 60
Mesopotamia 118, 119, 124–43, 158, 159, 160, 161, 210, 214
Southern 132, 158
Mesopotamian 5, 127, 129
Plain 120, 123, 125, 127, 130, 171
Mesopotamians 125
Mexicali Valley 72
Mexican 72
Mexico 24, 36, 67, 71, 72
Centre for Ecodevelopment 72
City 29
Gulf of 71
Northern 72
Meybeck, M. 68, 69
Micklin, P. P. 77
Middle Eastern and North African Countries (MENA) 45
Mississippi River 71
Mitchell, W. A. 117, 123, 131, 144, 157, 215
Mizyed, N. 37
Mongol 130
Morgan-Grenville, F. 48, 51, 219

Mori, S. 22
Morocco 177
Moscow 111
Moses 127
Murakami, M. 89, 92, 122
Musallam, R. 8, 9, 187, 191, 207
Muscat 217
Musiake, K. 89, 92
Muslim 3
Arabs 94
Muslims 2, 3
Mustafa, I. 95
Muynak 77
Myers, N. 75, 76, 165

Naff, T. 8, 20, 35, 37, 38, 61, 89, 90, 92, 93, 95, 99, 106, 113, 116, 121, 131, 135, 157, 212
Nahr al-Urdunn 85
Naess, A. 219
Naor, A. xi, 87, 112, 116
Narveson, J. 40
Nash, L. 68, 69
Nasser, Lake 19
National Water Carrier 24, 104, 105, 106
NATO (North Atlantic Treaty Organization) 136, 143
Nebopolassar 127
Nebuchadnezzar, King of Babylon 125, 127
Negev Desert 41, 66, 95, 105, 107
Netherlands, The 34, 35, 68, 207
Ngan, L. L. E. 94
Nicholson-Lord, D. 68
Nijim, B. K. 95, 207
Nile
River 4, 19, 85, 137, 195
basin 171
non-governmental organisations (NGOs) 15, 53
North American Water and Power Alliance (NAWAPA) 71
North Sea 68

Index

O'Riordan, T. 48
Ocalan, A. 155
OECD (Organization for Economic Cooperation and Development) 179
Ogallala aquifer 71
Oman 168, 175, 177, 181, 182, 187, 190, 195, 217
  Mountains 167, 168
  Southern 168
Omicinski, J. 190
Oppenheim, L. 36
Oriental 5
Orontes (Asi) River 90, 147, 149, 150, 153
Ostrom, E. 27
Ottoman 93
  Empire 130, 131, 133
  Turks 94, 130
Ozal, T. 154, 155, 212

Painton, F. 68, 69
Pakistan 20, 76, 195
Palestine 62, 93, 94, 95, 96, 97, 98, 99, 100, 215
  Arab refugees 98
  Northern 95
Palestinian 107, 111, 112
  National Liberation Movement (Al-Fatah) 105
  Territories, Occupied 89, 92
Palestinians 89, 91, 92, 98, 110, 112
Papp, D. S. 40
Paradise 2, 3
Pax Aquarum 117
Peace Pipeline 156, 195, 196, 197
Pearce, F. 107
Pentagon 116
Perera, J. 94
Peres, S. xi, 87, 106, 112, 116
Persian 1
  Gulf 132, 141, 153, 167, 168, 169, 181, 191, 195, 201
  Gulf Cooperation Council ([P]GCC) 183, 193, 198, 201

Gulf War (1990–1) 21, 136, 158, 164, 185, 191, 192, 201, 202
Persians 94
Peters, J. 217
Philistines 3
Poland 68, 69
Ponton, A. 48
Pontus 119
Pope, H. 121, 207
Porat, M. B. 108
Porritt, J. 52, 219
Postel, S. 9, 19, 21, 26, 29, 42, 43, 46, 51, 52, 53, 60, 62, 63, 67, 69, 71, 73, 74, 75, 76, 82, 92, 116, 117, 136, 169, 188, 190, 196, 219
Prophet, the 3
Punjab State 76

Qamhiyeh, A. A. 191
Qatar 142, 168, 181, 182, 190, 191, 195
  Solar Energy Research Station 191
Quingquan 74

Rajasthan 76
Ramadi 134
Rangeley, R. 76
Raphael, C. N. 188
Reagan, R. 69
Reisner, M. 70
Renner, M. G. 68
Repetto, R. 26
Restrepo, I. 72
Reuters 199
Rhine River 37, 68
Rio de Janerio 7
Rio Grande 36, 72
Riyadh 187
Romania 68
Romans 27, 94
Rome 60
Rub al Khali 167
Russia 66, 78

Sacramento River Delta 70

Safad 97
Sahara (Desert) 41, 61
Saharo-Arabian deserts 41
Saif, A.-A. M. 175
Samarkand 78
San Joaquin River 70
Sancton, T. A. 70
Santiago 29
Saqqa'in 173
Sardar Sarova Dam 75
Sargon II 127
Saudi Arabia 121, 165, 168, 169, 173, 180, 181, 182, 185, 186, 187, 188, 189, 190, 191, 193, 194, 195, 199, 200
  Central 168
  Eastern 168
  government 199
Saudis 201
Schoon, N. 82
Sennacherib 127
Senegal 19
  River 19, 37
Serageldin, I. 8, 21
Seven Wonders of the World 2
Sèvres, Treaty of (1920) 130
Seyhan River 195
Shabad, T. 76, 77
Shaibi, H. T. 188
Shapland, G. 140, 153, 160, 212
Sharett, M. 103, 107, 108
Sharon, A. 105
Shat al-Arab 122, 141, 149
Schulte-Wulwer-Leidig, A. 37
Siddartha Guatama (Buddha) 2
Sierra Nevada 70
Sikh Golden Temple 76
Sikhs 76
Silesia 69
Smith, C. D. 171, 179, 180
Smith, R. T. 32
Soroos, M. S. 62, 63, 64, 65
Soviet 79
  Union 24, 111
Spector, M. 78

Spooner, B. 171
Starke, L. 21, 52
Starr, J. R. 3, 8, 9, 80, 81, 106, 108, 109, 116, 121, 136, 145, 155, 156, 157, 164, 181, 184, 187, 191, 195, 201, 207, 212, 213
Stockholm International Peace Research Institute (SIPRI) 18
Stoll, D. C. 8, 106, 116, 121, 136, 164, 184, 187, 191, 195, 201
Stumm, W. 60, 66
Sudan 19, 41, 63, 195
Suez Affair 137
Suliman, M. 41
Sumerian Kingdom 125
Sun, P. 73, 82
Sutherlin, J. W. 106, 119
Swedish coast 69
Switzerland 68
Syr Darya River 76, 77, 79
Syria 20, 30, 63, 89, 90, 91, 95, 105, 116, 118, 120, 121, 122, 130, 132, 133, 135, 136, 137, 138, 140, 143, 144, 145, 146, 147, 148, 149, 150, 152, 153, 154, 155, 157, 158, 160, 161, 195, 217
  Iraqi crisis 138
  Northern 133
Syrian 105, 119, 122, 136, 146, 149, 152, 153, 157
  Iraqi 152
  Iraqi border 147
  prime minister 155

Tabqa (al 'Thawra) Dam 122, 138, 145, 146, 157
Tamil Nadu 75
Taurus 119
Tekeli, S. 149, 152
Temperley, T. G. 188, 190, 192
Temyr, S. 78
Tennessee Valley Water Authority 95
Texan 72

254 Index

Tharthar Canal 149, 158
Three Gorges Project 41
Third River 24, 158
Third World 20
Thomas, C. 36, 38, 56
Thompson, R. L. 18, 207
Tianjin 74
Tigris–Euphrates 131, 138
Tigris River 85, 119, 121, 122, 123, 126, 130, 132, 133, 134, 135, 139, 141, 144, 147, 148, 149, 154, 158, 161
Timberlake, L. 51, 72, 76
Time 70
Tinker, J. 72, 76
Tomanbay, M. 119
Trager, J. 211
Turkey 24, 41, 79, 116, 117, 118, 119, 121, 122, 130, 131, 132, 133, 134, 135, 136, 137, 138, 140, 142, 143, 144, 145, 146, 147, 148, 149, 150, 151, 152, 153, 154, 155, 156, 157, 158, 160, 161, 195, 196, 197, 212
  Asiatic 119
  Central 119
  Eastern 119
  Southern 147
Turkish
  army 154
  border 122
  government 130
  national economy 152
  politicians 152
  prime minister 155, 195
  Republic 130
  reservoir 146
  territory 121, 130, 134
  Syrian border 147
Turks 147, 149
  Ottoman 94, 130

United Arab Emirates (EAE) 168, 169, 180, 181, 182, 187, 189, 190, 191, 195
United Kingdom (UK) 68, 194
  Department of the Environment 68
United Nations (UN) 15, 38, 78, 95, 98, 149, 151, 158
  Conference on Environment and Development (UNCED) (1992) 7, 30, 31, 56
  Convention on the Law of the Non-navigational Uses of International Watercourses (UNCLNUIW) 37, 38, 39, 148, 150, 161
  Convention on the Law of the Sea (UNCLOS) 37
  Development Programme (UNDP) 79, 212
  Environment Programme (UNEP) 18, 19, 79, 212
  General Assembly (UNGA) 38, 39
  Water Conference in Mar del Plata (UNWC) 7, 63, 207
United States (of America; US/USA) 36, 41, 66, 69, 70, 71, 72, 116, 195, 212, 213, 217
  Army Corps of Engineers 116
  Department of Defence, Intelligence Agency 8
  Environmental Protection Agency (EPA) 70, 71
  history 69
  Mid-West 71
  military bases 143
  State Department 116
Ur 60
USSR 66, 135, 143
Utah 71
Uzbeck Academy of Science 78

Van Dijk, P. 47, 56, 218
Varanasi 2
Versailles Peace Conference 95
Viessman, W. 43, 45

Index 255

Vistula River 69
Vlachos, E. 7, 35, 63
Volga, River 68

Wagnick, K. 189
Wakil, M. 132, 147
Walker, T. 86, 136, 195, 197
War, Six Day (1967) 90, 105, 112
Warburg, P. 217
Ward, Christopher 52
Ward, Colin 47, 53
Warford, J. J. 26
Warner, F. 192
Warren, A. 179
Washington Times 41
Washington 195
Watson, A. 130
WE (*Water and Environment*) 195
WEI (*Water and Environment International*) 192
Weitz, J. 100
West 22
Bank 81, 89, 92, 106
Western 6, 51, 131, 132, 139, 158, 160, 211, 218, 219
Westlands 70
Westing, A. H. 18
Widstrand, C. 18, 21
Wilcox, W. 132
Williams, M. 51
Winpenny, J. 8, 23, 25-30, 32
Wittfogel, K. 3, 4, 5
Wolf, A. T. 9, 36, 37, 105, 113, 210

Wollman, S. 100
World Bank 6, 8, 22, 28, 30, 45, 50, 60, 79, 80, 107, 152, 153, 157, 212, 217
World Commission on Environment and Development (WCED) 50
World Health Organization (WHO) 69
Global Environmental Monitoring System 69
World Resources Institute 43, 44, 136
World War
First 94, 95, 98, 125, 130
Second 62, 132, 133, 134, 213
Worldwatch Institute 21
Wu, Y. A. 73, 74

Yarmouk River 89, 90, 91, 95
Yemen 167, 168, 175, 177, 180, 181, 182, 187, 190, 213
Yunus Khan, M. 76

Zagros 119
Zambesi basin 19
Zarour, H. 94, 105, 106, 110
Zarqa River 95
Zeitouni, N. 110
Zionism 86, 93, 94, 96, 102, 110, 112, 215
Zionist 61, 94, 95, 96-98, 100, 102, 110
Zionists 86, 94, 95, 103, 215
Labor 109